Coastal Research Library

Volume 26

Series Editor
Charles W. Finkl
Department of Geosciences
Florida Atlantic University
Boca Raton, FL, 33431, USA

The aim of this book series is to disseminate information to the coastal research community. The Series covers all aspects of coastal research including but not limited to relevant aspects of geological sciences, biology (incl. ecology and coastal marine ecosystems), geomorphology (physical geography), climate, littoral oceanography, coastal hydraulics, environmental (resource) management, engineering, and remote sensing. Policy, coastal law, and relevant issues such as conflict resolution and risk management would also be covered by the Series. The scope of the Series is broad and with a unique cross-disciplinary nature. The Series would tend to focus on topics that are of current interest and which carry some import as opposed to traditional titles that are esoteric and non-controversial. Monographs as well as contributed volumes are welcomed.

More information about this series at http://www.springer.com/series/8795

Nelson Rangel-Buitrago

Editor

Coastal Scenery

Evaluation and Management

 Springer

Editor
Nelson Rangel-Buitrago
Departamentos de Física y Biología
Universidad del Atlantico
Barranquilla, Atlantico, Colombia

ISSN 2211-0577 ISSN 2211-0585 (electronic)
Coastal Research Library
ISBN 978-3-030-07671-9 ISBN 978-3-319-78878-4 (eBook)
https://doi.org/10.1007/978-3-319-78878-4

Printed on acid-free paper

This Springer imprint is published by the registered company Springer International Publishing AG part of Springer Nature.
The registered company address is: Gewerbestrasse 11, 6330 Cham, Switzerland

Timeless scientific knowledge goes into the core of all natural phenomena and unites us across cultures, continents, and centuries. We dedicate this book to scientists who work for the benefit of mankind.

Foreword

The aesthetics of landscape, in this particular case coastal seascapes and shore, is difficult to tackle. The general subject area of "beauty" along the coast is considered in this volume in terms of methodological approaches to evaluation of coastal scenery. To some, this topic is esoteric and yet to others it has very specific practical applications to tourism. With increasing numbers of tourists visiting the coast, many shorefronts and beaches with densely populated hinterlands reel under the effects of overuse through density impacts associated with overpopulated beaches, for example, in China and elsewhere. The evaluative process is complicated by cultural norms of what is acceptable from environmental and socioeconomic points of view along different shores. Beach crowding in China or other Southeast Asian countries may locally be considered normal, whereas in other regions with more sparsely populated inland regions the beach space utilized by beach goers might appear to be under utilized. Cultural perception is conditioned by floating data points that make up impressions of what is beautiful or ugly. Whatever perceptual conditioning is extent along a particular coast, most beach users regard various aspects of pollution, on the shore or in the water, as unacceptable environmental conditions, but the range of acceptance is variable depending on the geographic region.

With this backdrop in mind, it is laudable that the authors of the seven chapters posit new methods for evaluating perceptions of beauty as part of scenic assessment techniques. Suggesting approaches to the study of landscape or scenery is based in the first instance on a review of prior efforts to describe the critical factors that are involved in the perception of what is beautiful or scenic. Determining the elements of the landscape that are involved here turns out to be a difficult task that is fraught with many difficulties.

Nonetheless, chapters in this book advance our understanding of how to better understand social mores associated with tourism and assess different approaches to evaluating and quantifying what is beautiful and desirable along the shore. As explained in this volume, it often is a delicate balance between sustaining scenic coastal landscapes and economics when the sheer volume of tourists strains coastal resources to the breaking point of land and water degradation. Such despoliation by the pressure of human bodies in limited spaces such as provided by beach berms is a very real problem in high user areas. Exploited here in various discussions are considerations of methods for achieving maximum use that provide reasonable economic returns without destroying the quality of that which was originally sought. Sustainability is a common buzzword these days, but the intensive use of many coastal environments is not possible, and limitations or restraints of human activities are required to conserve what is regarded as aesthetically desirable coastal segments.

Many advances in landscape assessment are supported by evaluation of maps, aerial photographs, and satellite images in digital formats that are amenable to study using GIS and computer modeling programs. These tools are critical resources that aid decision-making and natural resource planning at various levels where landscape characteristics, qualities, and influences on the landscape are recorded and evaluated. Landscape character assessment is, however, a complicated process that requires different levels of expertise in several endeavors such as geology, biology, environmental science, history, and socioeconomics. All of these approaches can be accommodated in modern GIS frameworks that help to classify landscape into areas of distinct visual and sensory character. These seascape character assessments are now maturing into codified approaches that help coastal managers suggest or adjust the fair and reasonable use of coastal resources for the benefit of stakeholders, tourists, and the environment itself that must be protected from uncontrolled human pressures. Readers of this book are thus introduced to the most up-to-date approaches for evaluating coastal scenery that include but which are not limited to descriptive inventories, public preference models, and quantitative holistic techniques that include psychophysical and surrogate component models.

This work is a requirement of coastal scenery evaluation because human pressure on coastal marine resources has reached the breaking point where once beautiful seascapes are now being loved to death. The virtual pressure of human body space along the coast and support resources for tourism (e.g., food and bathroom facilities, shops, motels and hotels, concession stands, travel agencies, emergency assistance, etc.) can adversely impact the coastal scene unless managerial positionalities are implemented according to the societal norms of the region. The authors provide here multi-pronged insights into the factors that make up coastal landscapes that are in turn evaluated in terms of their role in determining what constitutes scenic beauty.

Asheville, NC, USA   Charles W. Finkl

Preface

Countries with superb coastal scenery have an invaluable "plus effect" because the coast is an ideal place for tourism. It is necessary to remember that a well-managed coast is a perfect space in which social and economic activities can be done on a multiplicity of spatial and temporal scales obtaining an endless number of beneficiaries.

This book describes an easy to apply methodology to determine scenic value of the coast. As one of the most critical aspects of beach user choice, the determination of coastal area scenic quality is of primordial importance stated later.

This work is the first book to present a semiquantitative analysis of coastal scenery based on more than 4000 interviews about people's desired coastal preferences. Twenty-six parameters can be used to identify any coastal scene, which has been then subdivided into five attribute categories, weighted and subjected to fuzzy logic mathematics to obtain a decision number (D). This D number represents coastal scenery at that point, and five D classes are then presented (I – excellent to V – poor). Heritage areas and national parks should lie in Class I, which infers top scenic quality.

This book contains 7 chapters written by 6 authors from different parts of the world (Colombia, Italy, Malta, Turkey, and Wales), which between them, and over a time span of a decade or so, have assessed more than 952 global locations using the technique given in this book. One of the main aims of this method is to point out how scenic areas may be improved by judicious intervention relating to parameters, mainly anthropogenic, chosen for assessment.

The content of this book wants to open perspectives for analysis of the potential for coastal tourism development in natural areas together with landscape quality improvement in current coastal tourist developed areas. It will be a helpful tool for coastal lovers that include users, teachers, researchers, and managers.

Barranquilla, Colombia   Nelson Rangel-Buitrago

Acknowledgments

This book is a contribution from research groups: "Geology, Geophysics and Marine – Coastal Process," Universidad del Atlántico (Barranquilla, Colombia); "Coastal and Marine Research Group," University of Wales Trinity Saint David (Swansea, Wales, UK); "Department of Civil Engineering, Coastal Engineering Division," Middle East Technical University (Ankara, Turkey); "RNM-328," Universidad de Cadiz (Andalusia, Spain); "Euro-Mediterranean Centre on Insular Coastal Dynamics"; and "Institute of Earth Systems," University of Malta (Msida, Malta).

Thanks to British Council in Ankara, Turkey, and Valetta, Malta, for proving the financial support necessary to carry out the fieldwork of CSES throughout the British Council Project.

Thanks to Professors Andrew Short (University of Sydney) and Carlos Pereira da Silva (New University of Lisbon) for their useful comments and suggestions.

We also thank the following persons:

Nelson Rangel-Buitrago: Thanks to those individuals who have contributed in different ways during the development of this book: Adriana Gracia C., Magdalena Buitrago, Nelson Rangel Diaz, Fanny Rangel, William Neal, Jarbas and Carla Bonetti, Carolina Martinez, Manuel Contreras, and Victor de Jonge.

Allan Thomas Williams: Many people have contributed to my understanding of coasts but three in particular deserve mention. First, Prof. W.C. (Bill) Bradley (University of Colorado, USA), whose dynamic, brilliant, Geomorphology course ignited my love for beaches. This was followed by Prof. D.J. Dwyer (Hong Kong/Liverpool universities), who thankfully turned my attention from an economics/mathematics Ph.D. to a coastal one. Finally, Hilary, who allowed me to pursue coastal issues to my heart's content.

Ayşen Ergin: Thanks to those individuals who have contributed to motivation, and stimulating discussions cannot all be acknowledged properly here; however, very special thanks extended to Dr. Engin Karaesmen who has always with her interpretive capabilities and facility with mathematics contributed to this work in unique ways.

Giorgio Anfuso: Special thanks for field assistance go to Samanta da Costa Cristiano (for characterization of Brazilian beaches) and to Driss Nachite (for Morocco study case).

Enzo Pranzini: Thanks to Aysen, Allan, Anton, Giorgio and Nelson (yes, just my coauthors!) for having endured my constant criticism during the inspection to the beaches and the drafting of these texts.

Anton Micallef: I have been blessed with many friends who have inspired my love for the coast. I thank all of these, particularly those I have had the opportunity to work with in the field. I wish to acknowledge, in particular, my coauthors in this book for their dedication and encouragement and my parents, the latter, for gifting me as a child with the early books of Jacques Costeau who ignited my curiosity for the sea. Last but not least, I wish to thank Erika, my partner, for her constant support and patience. May others be as fortunate.

We also thank Petra van Steenbergen, Charles W. Fink (Coastal Research Library Series Editor), Shobha Karuppiah and Pavitra Arulmurugan, for their ongoing support and enthusiasm along this editorial project. Special Thanks to: Adriana Gracia C., Arthur and Ruth Maxwell who proof read this book.

Contents

Editor and Contributors

Editor

Nelson Rangel-Buitrago Departamentos de Física y Biología, Universidad del Atlántico, Barranquilla, Atlantico, Colombia

Contributors

Giorgio Anfuso Department of Earth Sciences, Faculty of Marine and Environmental Sciences, University of Cadiz, Cadiz, Spain

Ayşen Ergin Department of Civil Engineering, Coastal Engineering Division, Middle East Technical University, Ankara, Turkey

Anton Micallef Euro-Mediterranean Centre on Insular Coastal Dynamics, Institute of Earth Systems, University of Malta, Msida, Malta

Enzo Pranzini Department of Earth Sciences, University of Florence, Florence, Italy

Allan T. Williams Faculty of Architecture, Computing and Engineering, University of Wales, Swansea, Wales, UK

CICA NOVA, Nova Universidade de Lisboa, Lisbon, Portugal

Chapter 1
Coastal Scenery: An Introduction

Nelson Rangel-Buitrago, Allan T. Williams, Ayşen Ergin, Giorgio Anfuso, Anton Micallef, and Enzo Pranzini

> *'Mir hilft der Geist; auf einmal she'ich Rat.*
> *Und schreibe getrost. Im Anfang war die Tat.'*
> Goethe, Faust Part 1, lines 1236–7

Abstract Coastal tourism includes those recreational activities which involve travel away from one's place of residence which has as their host or focus the coastal zone. This industry necessarily depends on the coastal environment to attract tourists. Excellant scenery is maybe the prime factor considered by a potential tourist when is time to choose a coastal vacation destination. Coastal scenery management, a controlled tourism growth, an enhancing of the product, the constant upgrading of

N. Rangel-Buitrago (✉)
Departamentos de Física y Biologia, Facultad de Ciencias Básicas, Universidad del Atlántico, Barranquilla, Atlántico, Colombia
e-mail: nelsonrangel@mail.uniatlantico.edu.co

A. T. Williams
Faculty of Architecture, Computing and Engineering, University of Wales, Trinity Saint David, Swansea, Wales, UK

CICA NOVA, Nova Universidad de Lisboa, Lisbon, Portugal

A. Ergin
Department of Civil Engineering, Coastal Engineering Division, Middle East Technical University, Ankara, Turkey
e-mail: ergin@metu.edu.tr

G. Anfuso
Departamento de Ciencias de la Tierra, Facultad de Ciencias del Mar y Ambientales, Universidad de Cádiz, Puerto Real, Cádiz, Spain
e-mail: giorgio.anfuso@uca.es

A. Micallef
Euro-Mediterranean Centre on Insular Coastal Dynamics, Institute of Earth Systems, University of Malta, Msida, Malta
e-mail: anton.micallef@um.edu.mt

E. Pranzini
Department of Earth Sciences, University of Florence, Florence, Italy

the quality of offer and service, as well a diversified clientele, can be considered as critical points for an ideal tourism development that will satisfy both visitors and those whose livelihood depends on it.

1.1   Introduction

It is so small a thing to have enjoyed the sun. Mathew Arnold, Empedocles on Etna, 1, ii, 397

Coasts are the most dynamic and valued geomorphological features on the surface of the earth (Pilkey and Cooper 2014). They serve as home to a multitude of living organisms, including humans and are in continuous change due to a large variety of processes. From ancient times, coasts have played a significant role as a place for human settlement and economic development (Barragan and Andreis 2015). *'The coastline is of special importance'* (Steers 1944, 5), however, it is a very fragile environment easily affected by disordered infrastructures emplacement and activities, such as, industry, tourism, agriculture and fishing, amongst others.

During past years, there has been overdevelopment of many of these areas due to an unbridled pursuit for further economic benefits. This has led to an increase in environmental impacts due to processes that includes, amongst others, sand mining (Rangel-Buitrago et al. 2015a, b), beach pollution (Williams et al. 2013), and coastal armoring (Pranzini and Williams 2013).

The invaluable significance of coastal landscapes to society has long been recognized and is reflected by the plethora of existing protection status areas, such as, National Parks, Heritage Coasts, Wilderness Areas, Protected Landscapes and Areas of Outstanding Natural Beauty. However, despite the existence of these entities whose designations are strongly influenced by scenic beauty, scenic degradation globally greatly affects many coasts.

In the last few decades, the number of people able to visit coastal zones for recreational purposes has increased exponentially and correspondingly a popular desire to protect and conserve beautiful scenery has also risen over this period (UNEP 2009; Miller et al. 2010; UNWTO 2016). Frequently, coastal stakeholders and decision makers have been faced with a complicated question: **Should landscape development be impaired for the sake of conserving the natural scenery, or *vice versa*?** This can only be answered by determining what landscapes are favoured by society as a whole and this requires evaluation of the relative quality of coastal scenery by Governments in order that it can be compared to those of other landscapes and to the needs of other resource users. After all, *'coastal scenery is a resource, partly because of the economic value and partly because it is an accepted component of resource assessment programmes'* (Kaye and Alder 1999, 303–304). Evaluation of a coastal landscape is important as it provides measurement, description, and classification schemes (Dakin 2003; Ergin et al. 2004; Rangel-Buitrago et al. 2013), giving means by which scenery/amenities can be compared against other resource considerations (Ergin et al. 2006). It is a visual expression of the coast, and is a great resource that has not been analysed in detail on any scientific basis.

In addition, it can improve resource inventories, carrying capacity decision making, and can be included into Environmental Impact Assessments (Ergin et al. 2004). Coastal scenic evaluations allow managers to determine the relative attractiveness of locations so that informed decisions concerning improvements to the scenic quality of the landscape and their management may be made.

While this applies to all world landscapes, it is of particular importance to coastal scenery. Worldwide coastal scenery problems are further amplified by a tourist industry that is struggling to fill gaps left in the world economy by the decline of heavy industry and a rise in general affluence (Williams and Ergin 2004). The coastal tourist industry mainly depends on beaches to attract tourists (Botterill et al. 2000; White et al. 2010) and many diverse studies have shown that excellent coastal scenery is one of the major factors considered by tourists when choosing a beach vacation (Miller 1993; Unal and Williams 1999; Jędrzejczak 2004; Williams et al. 2016).

Scenery may be defined as *'the appearance of an area'* (Council of Europe 2000) and is a part of a coastal landscape inventory available for different coastal disciplines, such as, geography, geology, planning, etc. Likewise, coastal landscapes can be described as a littoral area, as perceived by humans, whose character results from the multiple interactions between natural and/or human factors (Council of Europe 2000).

Inside this book the reader can find an exhaustive review of existing scenery evaluation techniques, and can also obtain a novel methodology for coastal landscape evaluation, the Coastal Scenic Evaluation System (CSES) presented in Chap. 4, which is applied and presented by worldwide cases studies. However, it is salutary to note the words written over 70 years ago by a world leading coastal geographer that *'any assessment of coastal scenery is likely to meet with criticism'* (Steers 1944, 6). A series of recommendations is also given for adequate coastal scenery management.

Over a time span of a decade or so, the authors of this book have assessed more than 952 global locations by the technique given in Chaps. 4 and 5. Coastal/beach management, mainly driven by the tourist industry and appropriate government policies (designation of National Parks, Areas of Outstanding Natural Beauty, among others) has improved immensely and therefore the figures given here for coastal scenery may not represent the current situation. We urge readers to visit places mentioned in this book and assess their scenic value in order to realise an up to date figure for that particular location. One of the aims of the technique is to point out how scenic areas may be improved by judicious intervention relating to parameters, mainly anthropogenic, chosen for assessment.

The content of this book aims to open perspectives for analysis of the potential for coastal tourism development in natural areas and for scenic quality improvement in current coastal tourist developed areas. It will be a helpful tool for coastal lovers that includes users, teachers, researchers, and managers.

1.2   What Is Coastal Tourism?

Travel makes one modest. You see what a tiny place you occupy in the world. Gustave Flaubert

Despite it not being an easy task to provide an all-encompassing definition of tourism, this activity can be defined as the promotion and sale of the enjoyable and other features of a particular travel destination provision of facilities and services for pleasure travellers (UNWTO 2016). Tourism is an active, dynamic and competitive industry that demands the ability of continuous adaptation to customers' changing needs and desires, as client satisfaction, safety, and enjoyment is the main focus of the tourism business (WEF 2016).

The tourism industry is a complex activity, one that can be developed along distinct destination and environment lines, where a diversity of cultural, social, environmental and physical attractions exists. Increased tourism activity can kick start economic development within an area and consequently act as a catalyst for other related activities, which develop because of tourism. However, this can create negative impacts on sustainability, which in many cases are larger than the benefits that tourism brings (UNEP 2009; Holzner 2011).

Coastal Tourism, also known as **Sun, Sand, and Sea tourism (3S)** is based on a very particular resource conjunction along the interface between land and sea. This kind of activity offers amenities, such as, good weather conditions, water, beaches, scenic beauty, biodiversity, cultural and historical heritage, healthy food, and under optimal conditions an adequate infrastructure. With regard to visitor numbers and income generated over the past few years within the overall tourism sector, the 3S tourism market, is by far the most significant (UNTWO 2013). Among tourist destinations, coasts, are probably the prime factor favoured for visitor preferences (Lencek and Bosker 1998; Honey and Krantz 2007; Houston 2013). For example, the Mediterranean region is the world's leading tourist destination; almost one-third of global income from tourism is generated inside this region (UNTWO 2016).

Global studies reveal that the growth of 3S tourism has peaked in the last few decades (Miller 1993; Hall 2001; Moreno and Amelung 2009; Williams and Micallef 2009; Rangel-Buitrago et al. 2013). Its economic importance is unquestionable and its growth is strongly related to the natural physical characteristic e.g. scenery, as well as, socio-economic features of the receiving coastal area, such as, local community interests, health, political and security factors, together with the traditional models of tourism (UNEP 2009).

Some benefits that a well developed 3S tourism can generate include:

- Revenue generation and international receipts.
- Construction of Infrastructure and community facilities.
- Generation of new jobs and prosperity.
- Increasing awareness of the need for conservation.
- Production of sustainable community livelihoods.
- Investment in the environment and cultural heritage.
- Planning for potential end use (planning, environment tourism state and municipal authorities and academia).

One of the major issues that 3S tourism must resolve is the actual conflict between different benefits that provides this kind of activity and its effects on the coastal environment. These effects on coastal scenery can include:

Fig. 1.1 Examples of unsustainable developments along coastal areas with direct impacts over the environment. (**a**) Vegetation destruction by illegal buildings over the hills at Taganga Beach and (**b**) overdevelopment and armouring of buildings at Cartagena City, both located on the Caribbean Coast of Colombia. (**c**) Over population along Monte Carlo, Monaco and (**d**) Dune destruction and building construction in Mediq, Morroco

- Loss of habitat and biodiversity.
- Physical destruction and loss of amenity.
- Pollution.
- Property development patterns and motives.
- Resource consumption and competition.
- Limited community engagement and benefit.
- Seasonality and sensitivity of demand.

In some cases, coastal tourism activities are a process where any kind of decision is based on financial criteria, whilst the coastal environment is taken into account only when it is strictly necessary to minimize adverse effects that can threaten economic profit. A diversity of studies support this, for example, Lenceck and Bosker (1998), Defeo et al. (2007), Hughes and Duchain (2011), Pilkey and Cooper (2014).

This kind of process can lead to a chaotic and unsustainable development of coastal areas, which not only affects the environment but, in the medium term also severely impacts the different benefits of tourism since it modifies and destroys the sustenance of the tourism activity in these coastal areas: **Scenery** (Fig.1.1). The challenge here can be summarized in a very simple question that demands a smart answer: **How to develop a 3S tourism that will not minimize the quality of the natural resource and benefits to stakeholders?**

1.3 The Importance of Coastal Tourism (The 3S Market)

Twoflower was a tourist, the first ever seen on the disc world. Tourist, Rincewind had decided, meant 'idiot'. Terry Pratchett, The Color of Magic

Worldwide occur well dated antecedents that resemble current modern tourism, e.g. Herodotus and his attitudes to mobility, Cleopatra and Mark Antony with their own beach (Cleopatra's beach, Sidar Island, Turkey), Roman leisure travel along the Mediterranean coast (Greece – Egypt), and medieval pilgrimages. However, the well-known 3S tourism began under the Roman Empire with construction of the first hotels along the Italian peninsula (Baranowski et al. 2015). Through the following centuries, particularly from the eighteenth century onwards, 3S tourism was generally tied to therapeutic properties found at the coast. Sun, sea, and sand have throughout time provided the essential ingredients for coastal tourism until today and especially at the end of the twentieth century, which was clearly defined by the development of mass tourism (UNEP 2009).

Currently, tourism is one of the seven largest business sectors of the world economy (EEA 2006; UNTWO 2016). Their Gross Domestic Product contribution ranges from 2% for small scale tourism countries where tourism weighting can be significant, to more than 10% in countries where tourism is well developed (Briguglio 1995; Honey and Krants 2007). The industry generates one in twelve jobs globally, and between 35–40% of the world's export services (UNTWO 2016).

Since 1990, international tourism receipts have grown by 365%, moving from 271 to 1260 Billion US$ (UNTWO 2016). Despite there being no exact data on coastal tourism alone, the 3S industry is considered to be one of the largest-growing forms of travel in the last three decades and all 10 of the world's top destination countries in 2016 were countries with coastlines (Table 1.1). If the tourism industry were compared with a single country, it would have one of the world's major GDP's

Table 1.1 World's top 10 destination countries in 2016 with their related tourism statistics

Rank	Country	International tourist arrivals (× 1000)				International tourism receipts (US$ million)			
		2010	2013	2014	2015	2010	2013	2014	2015
1	France	77,648	83,634	83,701	84,452	47,013	56,562	5815	4592
2	United States	6001	69,995	75,022	7751	13,701	177,484	191,325	204,523
3	Spain	52,677	60,675	64,939	68,215	54,641	62,637	65,111	56,526
4	China	55,665	55,686	55,622	56,886	45,814	51,664	10,538	114,109
5	Italy	43,626	47,704	48,576	50,732	38,786	43,912	45,488	39,449
6	Turkey	31,364	37,795	39,811	39,478	22,585	27,997	29,552	26,616
7	Germany	26,875	31,545	33,005	34,972	34,679	41,279	43,321	36,867
8	United Kingdom	28,296	31,064	32,613	34,436	32,892	41,624	46,539	45,464
9	Mexico	23,290	24,151	29,346	32,093	11,992	13,949	16,208	17,734
10	Russian Federation	20,262	28,356	29,848	31,346	8831	11,988	11,759	8465

Source: UNWTO (2016)

and would use the same resources at the scale of a developed country, such as, Norway or Sweden. Worldwide, tourism ranks third as a global export category after oil and chemicals products, and ahead of food and automotive parts. In many countries tourism ranks as the first GDP contributor (e.g. Aruba, Malta, Spain). Tourism is now increasingly an essential component of economic diversification, both for advanced and emerging economies. Even, in the last few years, this industry has shown a capacity to compensate weaker revenues in many countries.

For example, The Caribbean, is a 3S tourist destination par excellence, where international tourist arrivals grew by 5.1 million (+27%) in the 2010–2015 period. This growth was driven by Cuba (+18%), Aruba (+14%), Barbados (+14%), Haiti (+11%), the Dominican Republic and Puerto Rico (both +9%). For South American countries, coastal tourism represents one of the most important economic activities. The development capacity of the Colombian 3S market appears to be almost limitless, with an increase of 593,000 international arrivals between 2010 and 2015 (MinCIT 2016; Rangel-Buitrago et al. 2018). The tourist industries rapid growth meant an increase of almost US$ 270–290 million per year for the Colombian GDP (Rangel-Buitrago et al. 2015a, b). The GDP relating to tourism activities (>US$ 3600 million in the balance of payments for travel/transportation), is the third highest source of foreign exchange after oil and coal, exceeding exports of coffee, and others products (ANATO 2015).

The Mediterranean coastal area tourist growth rate confirms its role as a primary bathing destination. Tourism, mainly 3S tourism, grew by 5% in 2015, with Spanish beaches, the sub-region's top destination and Europe's second largest, posting a 5% growth, and receiving a record 68 million international arrivals. Other coastal established destinations, Slovenia (+12%), Portugal (+10%), Croatia, Cyprus (both +9%), Greece (+7%) and Malta (+6%) reported strong results; emerging coastal destinations such as, Albania, Bosnia and Herzegovina, FYR Macedonia, Montenegro, and Serbia all reported double-digit increases (UNTWO 2016). The influence of 3S tourism is so high that peak population densities on the Mediterranean coast of Spain and France can reach 2400 people per square kilometer, this value is more than double those found in the winter season (EEA 2005).

The tourism industry has demonstrated a high ability to recover from short-term setbacks. The UNTWO (2013) recorded in 2009 a significant decline in global arrivals as a result of the world economic recession; 2010 recorded a growth. Despite possible future fluctuations, international visitors worldwide are expected to increase by 3.3% in the coming 13 years to reach 1.8 billion by 2030 (UNWTO 2013, Fig. 1.2). In addition arrivals in emerging coastal destinations (+4.5% a year) are expected to increase almost at twice the rate of those in advanced established economies (+2.3% a year).

The economic significance of coastal tourism is unquestionable, despite the fact that there has been some changes in coastal tourism demand. Rest and relaxation, based on sun, sea, and sand, remains the baseline, however, there has been a significant growth in the range of conditions that coastal tourists are seeking. These include:

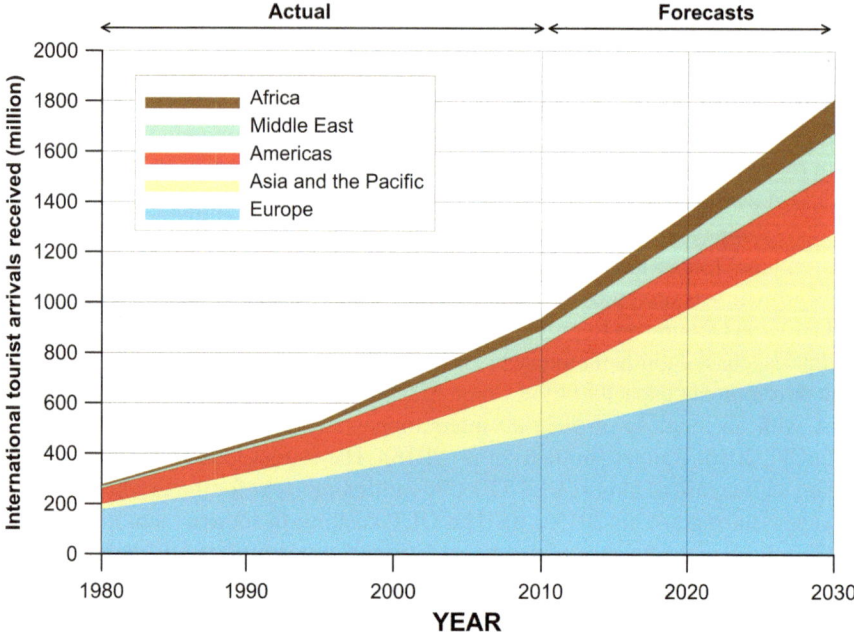

Fig. 1.2 Actual trends and forecasts of world tourism industry for the 1980–2030 period. (Data Source: UNWTO 2016)

- The quality of the environment and demand for clean and unspoiled locations.
- Availability of marine and coastal recreational activities.
- Excellent facilities.
- Nature and wildlife in optimal conditions.
- Presence of cultural attributes.
- Interest in combining coastal opportunities with other experiences through excursions.
- Use for the ecotourism.

The conditions mentioned above have a common characteristic, i.e. they are strongly related to **Coastal Scenery**.

1.4   Scenery as a Resource for Coastal Tourism

Mountains are the beginning and end of all scenery. John Ruskin, Modern Painters

Coastal development has always been strongly dependent on natural resource exploitation, among which is scenery. A coastal area in optimal conditions means millions of dollars profit (Clark 1996). Beach user's play a significant role in successful coastal tourism because they are the driving force and also the last receptor of the related economy. Many coastal countries must increase their tourist derived economic profits to be competitive and for this reason, it is critical to know beach user's preferences.

In regard to beach tourism, the above leads to an important question: **What do visitors rate highly?** The main answer is the bathing area, and from more than 4000 surveys carried out on beach user preferences worldwide, beach users were found to be primarily interested the in following five parameters the order of which can change according to beach typology (Williams 2011):

- Safety.
- Facilities.
- Water quality.
- Litter.
- Scenery.

It is Coastal Scenery, as shown by Ergin et al. (2004), Ullah et al. (2009), Rangel-Buitrago et al. (2013), Anfuso et al. (2014), among other authors, that is the goal of this book.

Since time immemorial, coastal scenery has always been a challenge to assess (see Chap. 3) and a source of inspiration for people and it can be considered as a resource for development in more than an economic sense. It can be exploited to generate employment and income, as well as developing the economy, but it is also a resource for other kinds of development (Hudson 1986). The impact of scenery on coastal tourism market is large and its influence in attracting visitors is much greater than shown in most existing beach award schemes (e.g. Blue Flag), as scenery is rarely mentioned (Van der Merwe et al. 2011; Lucrezi et al. 2015). But these schemes have been shown to not play a significant role in tourist's motivation to visit beaches (Mckenna et al. 2011).

Part of the lure of coastal scenery is the human image of a beautiful tropical beach. For many tourists this may be an undifferentiated image where in their minds, there exists a perception of a generalised amalgam of palm-fringed, white sand beaches washed by clear blue waters overlooked by luxuriantly forested mountains with cascading streams and exotic flowers and fruits, all under excellent weather conditions (Hudson 1986). After all, '*all our knowledge has its origins in our perceptions*' (da Vinci 1891). In some locations this idyllic panorama exists, however, scenic degradation is currently a big issue as in order to benefit from tourism arrivals, many coastal countries utilize growth policies which can devastate the coastal strip (Benoit and Comeau 2005; Martinez del Pozo and Anfuso 2008; Williams et al. 2018). For example, along the Mediterranean region, during the last two decades, 40% of the coastline has been lost to buildings, and by 2025, 50% would be irreversibly artificial (Benoit and Comeau 2005) A high density of buildings can currently be observed along several coastal areas worldwide, such as, Colombia, France, Italy, Mexico, Spain, where the built-up area exceeds 55% (EEA 2006, Fig. 1.3).

Benoit and Comeau (2005), showed that some 60% of locals interviewed in Italy and Spanish studies, commented not only upon poor planning on growth but also on

Fig. 1.3 Examples of changes in the density of constructed areas related to coastal tourism industry. Images (**a**) and (**b**) changes between 1975–2016 at El Puerto de Santa Maria, Cadiz (Spain), Images (**c**) and (**d**) are changes between 1954–2014 at Rodadero Beach, Magdalena (Colombia)

landscape degradation. For example, it is almost an axiom that when an island, such as, Malta, Aruba or Curacao can triple its summer population, some scenic degradation is bound to occur and this will ultimately affect tourism. Degradation affects a landscape immensely and scenery is therefore a vital component for the 3S market and drives the economy of many coastal countries, as beaches are under pressure from anthropogenic development and utilization (Ergin et al. 2006; Rangel-Buitrago et al. 2018). An adequate environmental management strategy, including effective physical planning, is necessary if coastal scenic quality is to be maintained. It is to be recalled that coastal scenery can be considered as a renewable resource because it can be sold to the tourism market and at the end, you can still have it…. of course, if it is well known how to preserve it!

1.5   Coastal Scenery Impact

Walking is a virtue, tourism is a deadly sin. Bruce Chatwin, What Am I Doing Here?

Countries with superb coastal scenery have an invaluable 'plus effect' because the coast is an ideal place for tourism. It is necessary to remember that a well-managed coast is a perfect space in which social and economic activities can be done in a multiplicity of spatial and temporal scales obtaining an endless number of beneficiaries.

In this regard, it is critical to differentiate between two concepts narrowly related, often erroneously seen as synonymous that can affect scenery and consequently any optimal coastal development: tourism growth and tourism development. Tourism growth can be measured in the number of arrivals, overnight stays, among other variables, which do not necessarily mean adequate economic prosperity (Ashley et al. 2007; UNEP 2009). Furthermore, tourism development refers to the increase in local income and employment, as well as, environmental benefits, thus implying the presence of development planning by the carrying capacity of the receiving environment (Neto 2003; UNEP 2009). Therefore coastal scenic deterioration can affect both growth and tourism development, and both can negatively influence coastal scenery.

In many cases, the urgency to obtain a faster economic profit derived from the 3S tourism has led to an uncontrolled and often elevated growth of this activity, which affects coastal scenery negatively. This was the case of Antigua and Barbuda (Kanji 2006), Belize (Diedrich 2010) and Jamaica (Henry 1988), that had to restart a wrong 3S tourism strategy based on an accelerated growth rate with the purpose of obtaining better results and preserving the coastal scenery resource. The above confirms that strengthening national economies by opting for quick economic profit from the tourism industry, is not necessarily the best alternative.

The observed growth of coastal tourism in the last decades is related to three main factors (EEA 2001; UNTWO 2016):

• Improvements in transportation systems.
• Increased personal incomes and leisure time.
• Knowledge of new world destinations due to improved communications.

Unfortunately, this growth has generated critical pressures on environmental and cultural resources of coastal areas and also has negatively impacted the economic, social and cultural patterns of some tourist destinations. However, with the increase in mass tourism in coastal zones, people involved in the investment and tourism management are increasingly aware that sustainability is strongly dependent upon the quality of these particularly fragile environments (UNEP 2009; Williams et al. 2016). As mentioned, 3S tourists want a diversity of experiences, in a lively and distinctive natural environment. Similarly, people living in coastal tourist destinations are increasingly aware of and concerned about their natural, historical and cultural heritage (CoastLearn 2009; Rangel-Buitrago et al. 2018).

An optimal coastal scenery management (Chap. 7), accompanied by a controlled tourism growth, an enhancing of the product, by upgrading the quality of offer and service, as a diversified clientele, can be considered as key points for an ideal tourism development satisfying both visitors and those whose livelihood depends on it. The above is the essence of "**Sustainable Tourism**." For example, during the last decades, the Rapa Nui population have enhanced coastal scenery by means a controlled tourism development, based on a constant upgrade of the quality of offer and service without leaving behind their own native traditions (Figs. 1.4 and 1.5).

To reduce tourism induced problems and secure both coastal resources and the sustainability of the tourism industry, significant attention must be paid to integra-

Fig. 1.4 Ahu Tongariki in Easter Island (Chile) an optimal coastal scenery management scenario

Fig. 1.5 Ahu Tongariki in Easter Island (Chile) an optimal coastal scenery management scenario

tion of 3S tourism into Integrated Coastal Zone Management (ICZM; UNEP 2009) now subsumed into Marine Spatial Planning (MSP).

For an optimal management of tourism development, it is of utmost importance to focus on appropriate planning of tourism growth taking into account scenery quality and the capacity of local systems (Chap. 6). Inside the ICZM/MSP framework, tourism is identified as one of the most important activities in coastal areas (Dodds and Kelman 2008; Williams and Micallef 2009; Haller et al. 2011). Diverse activities initiated by organizations e.g. the United Nations Environment Programme (UNEP), European Union (EU), European Environment Agency (EEA), United Nations World Tourism Organisation (UNWTO), all have highlighted the need to promote the implementation of actions for ICZM/MSP in different spatial – temporal scales. The ICZM/MSP approach provides a comprehensive set of measures associated with its development cycle, and today, these are widely applied worldwide. However, establishing its consistent and complete implementation within the 3S tourism sector remains a current challenge for all stakeholders. Coastal Scenery evaluation (Chap. 4) is a powerful tool that helps in sustainable development of tourism in coastal areas under the ICZM/MSP umbrella; it also provides invaluable baseline information and a scientific basis for any envisaged coastal development plan.

Reduction of tourism-induced issues and securing both coastal resources and sustainability of the tourism industry, demands greater attention to adequate management and better integration of tourism in coastal development. Conflicts and adverse impacts are due mainly to lack of understanding of coastal environments and weak, in some cases non-existent, management. A better knowledge of the physical environment of coastal areas, the identification of existing and potential uses and the development of integrated strategies and plans, offer the solution for a more socially and environmentally active development process (Ashe 2005; UNEP 2009).

1.6   Conclusions

'In a world of time and change there is no last chapter. The story never ends in a full stop.' A J Toynbee (1969) Experiences, 331

The world faces critical environmental issues and environment protection and conservation of natural resources at local and global scales are becoming more imperative. For this reason more than ever, diverse scientific disciplines must converge to provide necessary environmental solutions. Different management tools, such as Scenic Evaluation given in Chaps. 4 and 5, applied at an adequate stage of tourism development planning and within a clear regulatory and legislative framework, are a real guarantee of the sustainability and durability of tourism activity and its harmonious coexistence with other activities inside a well-preserved coastal environment.

References

ANATO (2015) Compendio de estadísticas turísticas de Colombia para el año 2015. ANATO, Bogota

Anfuso G, Lynch K, Williams AT, Perales JA, Pereira da Silva C, Nogueira Mendes R, Maanan M, Pretti C, Pranzini E, Winter C, Verdejo E, Ferreira M, Veiga J (2014) Comments on marine litter in oceans, seas and beaches: characteristics and impacts. Ann Mar Biol Res 2(1):1008–1114

Ashe JW (2005) Tourism investment as a tool for development and povertyreduction: the experience in small island Developing States (SIDS): trade and investment. The Commonwealth Finance Ministers, Bridgetown

Ashely C, De Brine P, Lehr A, Wilde H (2007) The role of the tourism sector in expanding economic opportunity. The Fellows of Harvard College, Cambridge

Baranowski S, Endy C, Hazbun W, Hom SM, Pirie G, Simmons T (2015) Tourism and empire. J Tour Hist 7(1–2):100–130

Barragan JM, Andreis M (2015) Analysis and trends of the world's coastal cities and agglomerations. Ocean Coast Manag 114:11–20

Benoit G, Comeau A (2005) A sustainable future for the mediterranean: the blue plan's environment and development outlook. Earthscan, London

Botterill D, Elwyn O, Emanuel L, Foster N, Gale T, Nelson C, Shelby M (2000) Perceptions from the periphery: the experience of wales. In: Frances B, Hall D (eds) Tourism in peripheral areas. Channel View Publications, Clevedon, pp 7–38

Briguglio L (1995) Small island developing states and their economic vulnerabilities. World Dev 23(9):1615–1632

Clark JR (1996) Coastal zone management handbook. CRC Press/Lewis Publishers, Boca Raton

CoastLearn (2009) Sustainable tourism. http://www.coastlearn.org

Council of Europe (2000) Landscape convention. Concil of Europe. Firenze

da Vinci L (1891) Codex Trivulzi, Castello Sforzesco. Pub. Luca Beltrami, Milan

Dakin S (2003) There's more to landscape than meets the eye: towards inclusive landscape assessment in resource and environmental management. Can Geogr 47(2):185–200

Defeo O, McLachlan A, Schoeman DS, Schlacher TA, Dugan J, Jones A, Lastra M, Scapini F (2007) Threats to sandy beach ecosystems: a review. Estuar Coast Shelf Sci 81:1–12

Diedrich A (2010) Cruise ship tourism in Belize: the implications of developing cruise ship tourism in an ecotourism destination. Ocean Coast Manag 3(5–6):234–244

Dodds R, Kelman I (2008) How climate change is considered in sustainable tourism policies: a case of the Mediterranean islands of Malta and Mallorca. Tour Rev Int 12:57–70

EEA (2001) Environmental signals for 2001. Office for Official Publications of the European Communities, Copenhagen

EEA (2005) The European environment – state and outlook. Office for Official Publications of the European Communities, Copenhagen

EEA (2006) The changing faces of Europe's coastal areas. Office for Official Publications of the European Communities, Copenhagen

Ergin A, Karaesmen E, Micallef A, Williams AT (2004) A new methodology for evaluating coastal scenery: fuzzy logic systems. Area 36:367–386

Ergin A, Williams AT, Micallef A (2006) Coastal scenery: appreciation and evaluation. J Coast Res 22(4):958–964

Hall MC (2001) Trends in ocean and coastal tourism: the end of the last frontier? Ocean Coast Manag 44(9–10):601–618

Haller I, Stybel N, Schumacher S, Mossbauer M (2011) Will beaches be enough? Future changes for coastal tourism at the German Baltic Sea. J Coast Res 61(61):70–80

Henry B (1988) The environmental impact of tourism in Jamaica. Tour Rev 43(2):16–19

Holzner M (2011) Tourism and economic development: the beach disease? Tour Manag 32:923–933

Honey M, Krants D (2007) Global trends in coastal tourism. Marine Program World Wildlife Fund, Washington, DC

Houston JR (2013) The economic value of beaches – a 2013 update. Shore Beach 81(1):3–11

Hudson BJ (1986) Landscape as resource for national development: a Caribbean view. Geography 71(2):116–121

Hughes Z, Duchain H (2011) Tourism and climate impact on the North American Eastern seaboard. In: Jones A, Phillips MR (eds) Disappearing destinations: climate change and future challenges for coastal tourism. CABI, Oxford, pp 161–176

Jędrzejczak MF (2004) The modern tourist's perception of the beach: is the sandy beach a place of conflict between tourism and biodiversity? In: Schernewski G, Löser N (eds) Managing the Baltic sea. Coastline reports. EUCC, Berlin, pp 109–119

Kanji F (2006) A global perspective on the challenges of coastal tourism. Coastal Development Centre, Bangkok

Kaye R, Alder J (1999) Coastal planning and management. E & FN, Spon, New York

Lencek L, Bosker G (1998) The beach: the history of paradise on earth. Viking Penguin, New York

Lucrezi S, Saayman M, Van der Merwe P (2015) Managing beaches and beachgoers: lessons from and for the Blue Flag award. Tour Manag 48:211–230

Martinez del Pozo JA, Anfuso G (2008) Spatial approach to medium-term coastal evolution in South Sicily (Italy): implications for coastal erosion management. J Coast Res 24(1):33–42

McKenna J, Williams AT, Cooper JAG (2011) Blue flag or Red Herring: do beach awards encourage the public to visit beaches? Tour Manag 32:576–588

Miller M (1993) The rise of coastal and marine tourism. Ocean Coast Manag 20(3):181–199

Miller G, Rathouse K, Scarles C, Holmes K, Tribe J (2010) Public understanding of sustainable tourism. Ann Tour Res 37:627–645

MinCIT (2016) Estadísticas de turismo en Colombia para el año 2015. FONTUR, Bogotá

Moreno A, Amelung B (2009) Impacts of climate change in tourism in Europe. JRC European Commission, Sevilla

Neto F (2003) A new approach to sustainable tourism development: moving beyond environmental protection. Department of Economic and Social Affairs United Nations, New York

Pilkey O, Cooper A (2014) The last beach. Duke University Press, Durham

Pranzini E, Williams AT (2013) Coastal erosion and protection in Europe. Routledge/Earthscan, London

Rangel-Buitrago N, Anfuso G, Correa I, Ergin A, Williams AT (2013) Assessing and managing scenery of the Caribbean Coast of Colombia. Tour Manag 35:41–58

Rangel-Buitrago N, Anfuso G, Ergin A, Williams AT (2015a) Assessing and managing the coastal scenery: blue solutions from Latin America and the wider Caribbean. GTZ, Berlin

Rangel-Buitrago N, Anfuso G, William AT (2015b) Coastal erosion along the Caribbean coast of Colombia: magnitudes, causes and management. Ocean Coast Manag 114:129–144

Rangel-Buitrago N, Williams AT, Anfuso G (2018) Hard protection structures as a principal coastal erosion management strategy along the Caribbean coast of Colombia. A chronicle of pitfalls. Ocean Coast Manag 156:43–58. https://doi.org/10.1016/j.ocecoaman.2017.04.006

Steers JA (1944) Coastal preservation and plannng. Geogr J 104:7–27

Toynbee AJ (1969) Experiences. Oxford University Press, Oxford

Ullah Z, Johnson D, Micallef A, Williams AT (2009) From the Mediterranean to Pakistan and back – coastal scenic assessment for tourism development in Pakistan. J Coast Conv Manag 14(4):285–293

Unal O, Williams AT (1999) Beach visits and willingness to pay: Cesme peninsula, Turkey. In: Ozhan E (ed) Medcoast 99 –land ocean interactions: monitoring coastal ecosystems. MEDCOAST, Ankara, pp 1149–1162

UNEP (2009) Sustainable coastal tourism – an integrated planning management approach. United Nations Environment Programme, Milan

UNWTO (2013) World tourism highlights 2013. UNTWO, Madrid

UNWTO (2016) World tourism highlights 2016. UNTWO, Madrid

Van der Merwe P, Slabbert E, Saayman M (2011) Travel motivations of tourists to selected marine destinations. Int J Tour Res 13:457–467

WEF (2016) Travel and tourism competitiveness report 2015. World Economic Forum, Geneva

White MP, Smith A, Humphryes K, Pahl S, Cracknell D, Depledge M (2010) Blue space: the importance of water for preference, affect, and restorativeness ratings of natural and built scenes. J Environ Psychol 30:482–493

Williams AT (2011) Definitions and typologies of coastal tourism beach destinations. In: Jones AL, Phillips MR (eds) Disappearing destinations: climate change and future challenges for coastal tourism. CABI, Oxford, pp 47–66

Williams AT, Ergin A (2004) Heritage coasts in Wales, UK. In: Micallef A, Vassallo A (eds) Proceedings of the first international conference on the management of coastal recreational resources – beaches, Yacht Marinas and Ecotourism, Malta. Euro-Mediterranean Centre on Insular Coastal Dynamics (ICoD), Gozo, pp 219–227

Williams AT, Micallef A (2009) Beach management principles and practice. Earthscan, London

Williams AT, Pond K, Ergin A, Cullis MJ (2013) The hazards of beach litter. In: Finkl C (ed) Coastal hazards. Springer, New York, pp 753–780

Williams AT, Rangel-Buitrago N, Anfuso G, Cervantes O, Botero C (2016) Litter impacts on scenery and tourism on the Colombian north Caribbean coast. Tour Manag 55:209–224

Williams AT, Rangel-Buitrago N, Pranzini E, Anfuso G (2018) The management of coastal erosion. Ocean Coast Manag 156:4–20. https://doi.org/10.1016/j.ocecoaman.2017.03.022

Chapter 2
The Concept of Scenic Beauty in a Landscape

Allan T. Williams

'A thing of beauty is a joy for ever/its loveliness increases.'

(Keats 1918)

Abstract The concept of beauty has for many centuries been considered and debated by philosophers, e.g. Kant, Wittgenstein, Hume and Locke. It is an ephemeral word that conjures up different meanings in people's minds alongside its counterpart ugliness. When the term is applied to coastal scenery the spectrum of measuring beautiful scenery has been a task that has occupied geographers, planners, etc. for at least a century. Beautiful scenery is a prime criterion for areas, such as, National Parks, Heritage Coasts, Areas of Outstanding Natural Beauty, but how is it assessed? Quality in a landscape is intrinsic in the physical quality of the area and is also a product of the mind of the observer, *i.e.* the scene looked at by an observer interacts with his/her perception of it to make a value judgment. If this is high, then the scene has beauty. Any landscape consists of historical, social and aesthetic aspects and this chapter concerns itself with these parameters, especially the visual aspect of the latter.

A. T. Williams (✉)
Faculty of Architecture, Computing and Engineering, University of Wales, Trinity Saint David, Swansea, Wales, UK

CICA NOVA, Nova Universidade de Lisboa, Lisbon, Portugal

© Springer International Publishing AG, part of Springer Nature 2019
N. Rangel-Buitrago (ed.), *Coastal Scenery*, Coastal Research Library 26,
https://doi.org/10.1007/978-3-319-78878-4_2

2.1 Introduction

The love of nature is the same passion as the love of the magnificent, the sublime, and the beautiful. Sir Humphrey Davy (1840, 307–308)

Beautiful scenery is a very important component for coastal tourism and is a major driver of the economy of many coastal countries where these areas are frequently under pressure, mainly from anthropogenic development and utilisation (Ergin et al. 2006). Beautiful coastal scenery is but one of five main parameters that beach users desire, the others being: safety, facilities, water quality and no litter (Williams and Micallef 2009). Scenery is just part of a coastal landscape inventory available for different coastal zone disciplines and its evaluation is not only an important instrument simply for tourists, but also for those of coastal preservation, protection and development, as evaluation outcomes provide a scientific basis for any envisaged management plan. A basic difficulty in assessing scenic – relating to, '*scenery, having beautiful or remarkable scenery,*' (Chambers 2000) quality is that of definition, as it is an abstract concept that is greatly confused by semantic difficulties, misunderstandings and controversies. Hull and Revell (1989) classified a scene as a landscape subset viewed from one location (vantage point) looking in one direction.

Some definitions of scenery are:

- '*the appearance of land*' (CC 1987, section 3.2).
- '*the aggregate of picturesque features in a landscape*' (Oxford dictionaries 2016).
- '*views presented by natural features, especially when picturesque*' (Cassell dictionary 1997).
- '*views of beautiful, picturesque or impressive country*' (Chambers dictionary 2000).
- '*the appearance of an area*' (CoE 2000, 4).

Kaye and Alder (1999, 302–303) argued that scenery, '*is a resource, partly because of the economic value and partly because it is an accepted component of resource assessment programmes*' and is therefore a variable that should be considered in land use decisions (Dearden 1985). If scenery is a resource, measurement of this resource, which is strongly rooted in the man-environment tradition, has taxed the brains of many eminent scientists. It is a veritable palimpsest and managers need to attempt evaluation of scenic resources in an objective and quantitative manner. Scenery is a major component not only of coastal, but of any landscape, as Sauer's (1938) benchmark paper pointed out and it served as a key concept for the Berkley school of American cultural geography.

Amir and Gidalizon (1990) argued that the above focuses upon an environment's visual properties/characteristics including natural and anthropogenic elements plus physical and biological resources. Non-visual biological functions, cultural/historical values, wildlife and endangered species, wilderness value, opportunities for recreation activities and a large array of tastes, smells and feelings were not included.

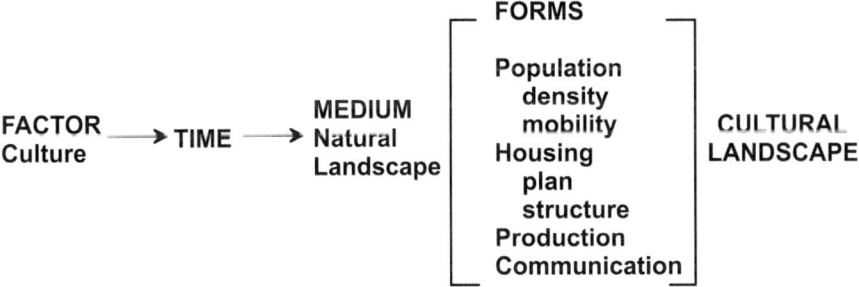

Fig. 2.1 Man's activities on the landscape (Sauer 1938, 318)

Olwig (2003) has given an excellent account of the above. Environmental anthropogenic and physical parameters are exemplified in Katsushika Hokusai's iconic 'Great Wave off Kanagawa' a 39 cm × 26 cm woodblock print made between 1830 and 1833, which seemingly symbolised nature's irresistible force, the weakness of human beings and the unpredictability of life.

The definition followed in this chapter is that of Sauer (1938 319), who defined landscape as, *'an area made up of a distinct association of forms, both physical and cultural.... The content of landscape is found... in the physical qualities of the area that are significant to man and in the form of his use of the area, in facts of physical background and facts of human culture,'* *i.e.* a cultural landscape is fashioned from the natural physical one by the activities of the cultural group (Fig. 2.1).

Sauer (1938) inferred that landscape was concerned with phenomenological shape/form and the way in which it shaped the world, a concept that Hartshorne (1939) disputed, who argued that landscape studies were of little value and had no fixed defined meaning. Lowenthal (1972), in a seminal paper on perceptual geography, challenged and changed this viewpoint by expanding Sauer's (1938) approach showing that landscape was the fulcrum of environmental concern and heritage and was a just phenomenon for epistemological inquiry. Lowenthal (1972) was fascinated by the many ways whereby people, societies (farmers, fisherman, poets, painters, business executives, etc.) knew and viewed landscapes and how they shaped the environment, *i.e.* from the pictures in one's head and the shape/appearance of the natural landscape. Cosgrove and Daniels (1988, 8) built on this viewpoint of landscape as *'a cultural image, a pictorial way of representing or symbolising surroundings,'* which made the world outside a minor component of landscape. This viewpoint has been criticised by Ingold (2000), who rejected the concept of inner and outer worlds, *i.e.* mind and matter. He was influenced by Heidegger's philosophy in that the material landscape is where man exists as a living being and who shapes the environment.

The term Landscape originally derived from Dutch (landscap) introduced into England in the sixteenth century by art dealers and art critics who translated it as meaning *'.... a picture of (Dutch) inland scenery'*, (Penning-Rowsell and Lowenthal 1986, 5). From the mid eighteenth century it became a popular talking point and an

Fig. 2.2 Examples of landscape and their key elements. (**a**) coastal arch at the Dorset coast, UK, (**b**) rock stacks at Reynisdranger coast at Iceland, (**c**) Maya ruins at Tulum coast, Mexico, (**d**) Canon at La Preneuse beach, Mauritius, (**e**) Trafalgar lighthouse, Barbate, Spain

enthusiasm for mountain scenery became fashionable amongst a public who could afford to travel (Cox 1988).

Any landscape comprises three elements: Social, Aesthetic, Historical (Fig. 2.2). The aesthetic is mainly the visual parameter the one mainly concerned with scenic grandeur exemplified by Jovaisa's (2015, 213) comment regarding scenery in eastern Cuba as *'the most beautiful landscape human eyes had ever seen.'* Even the great artist Courbet argued that beauty was only to be found in nature, as *'beauty provided by nature is superior to all artists' compositions'* (Barnes 2015, 64), but how does one determine this?

This differed from Hogarth's (1753) viewpoint of the S shaped sensuous line of beauty of visual art *ie.* the form of the object – his 'line of beauty' and 'line of grace.' Hogarth (1753) developed a universal set of aesthetic principles, *i.e.* fitness, variety, regularity, simplicity, intricacy and quantity (greatness, which had an aesthetic effect on the beholder) – the 'wow' factor of number, weight and measure according to mathematical precision. Therefore it contains the visual (extent, scale, continuity, colour, diversity, views, etc.); the landscape elements of geology, relief, soil, vegetation, ecology, archaeology, etc. as well as the other senses, such as, smell, taste, together with the feelings evoked in the observer, e.g. joy, comfort, awe.

Aspects of the above concepts have frequently appeared in many scientific papers/books referring to classical definitions of beauty, e.g. harmony, simplicity, proportion (Eco 2016), but even within western philosophy these meanings have changed during historic periods, *i.e. 'Renaissance man saw the proportions of cathedrals as barbarous'* (Eco 2011, 10). But can beauty only exist in harmony? Jackson Pollack paintings can evoke a feeling of the beauty that is perceived and this is in paintings that exhibit discord and dissonance. On the other hand, Constable's paintings epitomise harmony with nature. Perhaps the beauty of a messy complexity is of a higher order than that of harmony, an argument put forward by Weinberg (1993) in discussing 17 varieties of scientific beauty, which generally were subordinate to scientific usefulness and truth.

Eckbo (1975, 31) suggested four major dimensions to the landscape term. *'The physical landscape.... A 4d sequential pattern of earth, rock, water, plants manmade structures, air, weather, light and energy.... The social landscape expresses the local, regional, national and world wide relations among the people with whom we live. The economic landscape determines how well we live. The cultural landscape embodies the creative contribution of our times to world history.'* Tudor (2014) summed this up extremely neatly (Fig. 2.3).

Following Sauer (1938), Fairbrother (1970), suggested that the nature of landscape could be summed up in a conceptual Eq. (2.1):

$$LANDSCAPE + HABITAT = MAN \qquad (2.1)$$

If either man or the habitat changes so would the resulting landscape and this equation can never be stable due to the fact that man is constantly changing the habitat.

Landscape quality assessment is based on two contrasting paradigms, one regards quality as intrinsically inherent in the physical landscape, assessed by applying certain criteria subjectively presented as objectivity; the other regards quality as a product of the mind and the eye of the beholder and assessed by psychological methods with an objective evaluation of subjectivity *i.e.* it is a human construct. Respectively, these can be termed the objectivist (physical) paradigm as advocated by, for example, Plato (Waterfield 1993), Leopold (1969), Linton (1968, 1982), and the subjectivist (psychological) paradigm, as followed by Locke (1816), Hume (1757), Kant (1790), Buhyoff and Arndt (1981), Buhyoff et al. (1986), Kaplan and Kaplan (1989a, b), Purcell (1992), Purcell and Lamb, (1998). Both can-

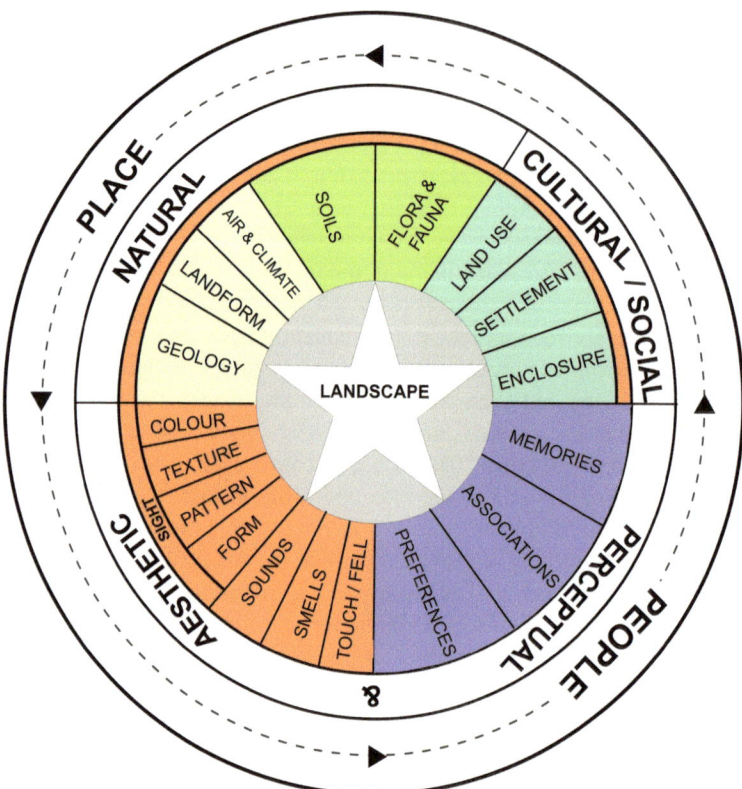

Fig. 2.3 Landscape elements in depth (Tudor 2014)

not be correct due to their contrasting underlying premises. '*The paradox is that in common usage, the landscape is taken to be beautiful but in actuality this beauty is literally a figment of the imagination, a product of the viewer's own cultural, social and psychological constitution*' (Lothian 1999, 180). Therefore do we see beauty or are intellectual as well as sensual factors involved? Culture plays a major part, as it has given mankind an aptitude for seeing a certain kind of beauty, but it still remains elusive, fugitive and evanescent.

2.2   Scenic Beauty

> *I look for the lover of nature in the artist and for the artist in the scientist. We belong together*. Willstätter (1965, 395)

'*Beauty has never been absolute and immutable but has taken on different aspects depending on the historical period and the country*' (Eco 2016, 14) and one definition is, '*the quality that gives pleasure to the sight or aesthetic pleasure generally*'

(Chambers dictionary 2000); or, *'that quality or assemblage of qualities which gives the eye or the other senses intense pleasure'* (Cassell dictionary 1997). It is a hard task to try to quantify/semi-quantify visual beauty (Fig. 2.2), but as Walt Disney stated, *'It's kind of fun to do the impossible,'* quoted in Skylar (1967). Porteous (1985) showed that up to 90% of our perceptual intake is visual and much of the rest is auditory and tactile. The smell of new mown hay, watching a murmuration of starlings swoop though the sky, listening to the music emanating from a blackbird's throat, all are primary sensations and have meaning, i.e. one does not have to see in order to touch.

It is impossible to quantify the non-visual aesthetic aspects, as they are spasmodic, temporary phenomena. With respect to sound, Porteous and Maskin (1985) showed that natural sounds were the most appreciated e.g. sound of the surf; smell, Eisner (1999, 451) joked that, *'I'm essentially a nose with a human being attached'* (Levi and Regge 1989, 62) wrote that, *'I'm very glad that I educated my nose'*. The viewpoint that humans have a poor smell sense is a myth dating from the nineteenth century (McGann 2017). The human olfactory bulb across 24 mammalian species is comparatively similar but proportionately smaller containing fewer functional olfactory receptor genes than rodents, but the neuron numbers are comparable. As *Homo sapiens* evolved the olfactory bulb did not become smaller, the rest of the brain became larger. Over appropriate odour ranges humans can outperform laboratory rodents and dogs in detecting some odours, but are less sensitive to others (Stern 2017; www.science.sciencemag). Taste seemingly has little influence on scenic beauty, although a salt laden spray can invoke feelings, but one cannot 'touch' scenery. One may touch the ground or a tree, but not the whole grandiose sweep of a scene. Movement can affect the impression of beauty: waves are beautiful not only for their shape, but also because of their propagation. Tree fronds and high grass moving under a strong wind are much more 'beautiful' than in a static shape. All however play just a small part in adding/detracting from the appreciation of scenic beauty (Morisawa 1971) and they are also unmeasurable.

As regards the historical (Fig. 2.2), any landscape evolves through time – from verdant fields to the dark satanic mills of William Blake – back to fields/towns (Ingold 1993). The effect of *milieu* on man came during the UK Romantic period (late eighteenth century to mid nineteenth century) which *'opened people's eyes to the sublime beauty of landscape, and, in due course, to the grim symbolism of cityscapes in the Industrial Age.'* (Lodge 1992, 58).

As for the cultural landscape (Fig. 2.2), the Aborigine viewpoint best expresses this concept, even though, as Winton (2016, 13) commented, *'Australia is still a place where there is more landscape than culture.... Everything we do it still overborne and underwritten by the seething tumult of nature'*; where a great deal of *'Aboriginal culture is arcane and dizzyingly complex to the outsider'* (Winton 2016, 189). Mowaljarlai (1995) in a radio broadcast expressed this most succinctly: *'We cry for you because you haven't got any meaning of culture in this country.... It's the culture of blood of this country, of Aboriginal groups, of the ecology of the land itself.'* (Wright 2005, 68) corroborates these views, writing ...*'So I came to Uluru six years ago. I sat down with my grandparents and told them about my piece of paper, and they said 'that can't find you water, food. Now you have to come with us to get your cultural papers from the landscape.'* This correlates with Duncan and Duncan's (1992) findings of the semiologist Roland Barthes, who postulated the importance of signs in the cultural landscape, as did Appleton (1996). Cosgrove and Daniels (1988) have given a sound account of the iconography involved.

If one of the main tenets of coastal tourism is the desire for *'scenic beauty'*, what is the etymology of the words, remembering that, *'one cannot speak of beauty, one has to experience it,'* (Böhme 2010, 23) The word beauty is an abstract quality, as it is not a scientific concept, e.g. mass, time, (Langer 1953, 5). Interestingly it was once believed, *'that there was enough beauty for everyone, so there was no need to conserve it'* (Bufford 1973, 1438). Beauty is popularly considered to be the central dimension of one's aesthetic relationship with the environment representing the highest common factor of quality derived from one's focused acts of environmental appreciation. From a theological-metaphysical viewpoint, the universe is beautiful, as it was the work of the divine. Newton (1966, 24) wrote: *'except within the vaguest limits, beauty cannot be described: therefore it cannot be defined. It cannot be measured either in quantity or quality; therefore it cannot be made into the basis of a science. It has always proved impregnable to the frontal attacks of the aestheticians'*. Much progress has been made since this remark was made, as is evidenced in this book.

Schopenhauer thought that the feeling of the sublime was identical with the feeling of the beautiful, the sublime feeling being one of delight and not pleasure, this being negative pleasure (Burke 1757/2008) *i.e.* removal of pain/danger (Vandenabeele 2003). Tatarkiewicz (1972) gave a superb erudite exposition on this matter, tracing the concept throughout history. Defining beauty is difficult and virtually synonymous with aesthetics, the name given for the philosophical study of art and natural beauty (Guyer 2004), first introduced into the literature by Baumgarten (1735), but in essence it is the branch of philosophy dealing with the basic concepts of thinking about objects of beauty (Shuttleworth 1983). However, the contemporary use of the term 'aesthetics' originated in the *'Critique of Judgment'* by Kant (1790). Foster (1991) reinterpreted Kant and Schopenhauer's views by examining the contribution of art and science to environmental aesthetics, suggesting an aesthetic framework based upon a philosophical fundament for the natural environment. Aesthetic appreciation and judgement was *'neither bound to art or science'*, but to *'perceptual features, aesthetic properties and descriptive qualities'* (Foster 1991, 222), rejecting the cognitive influence on landscape preference views of, e.g. Penning-Rowsell and Lowenthal's (1986), arguing for the importance of the sensuous experience of nature (Foster 1991, 94). The case for a multi-sensuous aesthetic appreciation of nature, especially ecology, was made, which makes for *'a unique sense of place'* (Foster 1991 182).

Baumgarten's (1735) viewpoint designated the concept of beauty as gathered through sensation resulting in perfection, *i.e.* the quality present in a thing/person that gives intense pleasure or deep satisfaction to the mind, whether arising from sensory manifestations, e.g. shape, colour, sound or other matters. Baumgarten (1735) took the object of beauty beyond the limitation of art, defining it as the science of perception, as prior to this the whole emphasis on beauty revolved around religion. For example, Bishop Grosseteste (c.1168–1253) argued that there was no beauty in the world apart from God (McEvoy 2000); and the concept of harmony, e.g. St. Augustine, who coined the venerable formulation for beauty: measure, shape and order, *'only beauty pleases; and in beauty, shapes; in shapes, proportions; and in proportions, numbers'* (St. Augustine (in); Tatarkiewicz 1972, 168). Tatarkiewicz (1972, 167) wrote that Aristotle adhered to the same view and asserted that *'beauty consists in magnitude and ordered arrangement and that the main forms of beauty are order, proportion, and definiteness.'* This is encapsulated in the ecclesiastical realm, where edifices range from the gothic Notre Dame cathedral, Paris, through the majesty of the much later Sacre Coeur church, Paris, to simpler churches, such as, the Church of the Holy Trinity, Stratford-upon-Avon, UK, where Shakespeare is buried, to an even simpler church at Oare, Exmoor, UK, immortalised in the book, Lorna Doone (Blackmore 1993).

It is important to note that beauty especially, natural beauty evokes aesthetic responses, which form an important part in any hierarchy of values and represents a source of self-actualisation for which man strives once basic biological/psychological needs are satisfied (Maslow 1968). It is best understood as an experiential product of the interaction of man and his surroundings (Zube and Sell 1986). Boulton (1958, 112), quoting Burke, argued that *'beauty is, for the greater part, some quality in bodies, acting mechanically upon the human mind by the intervention of the senses.'* Landforms and land use patterns are important information sources, as together with the individual's landscape range of experiences, they shape individual perceptions (Fig. 2.4 and Chap. 3).

However, no two individuals see the world in the same light, each brain is unique and experience influences perception and no two people can ever have exactly the same life experiences. In Lowenthal's (1972) seminal paper the two key terms on which perception depends relate to 'experience' and 'imagination.' Not everyone can agree as to what is beautiful, but most seem able to recognise the worst aspects of ugliness, the assumption being that the two are not reversible 'bi-polar opposites' (Kates 1966, 25). Therefore beauty is elusive, personal, subjective and often related to emotion and, as Dewey (1929) maintained, beauty can be discovered in the relationship between the individual and his environment, what he termed 'experience', *i.e.* it is a two way contact.

In 1781 the age old arguments concerning religion and science came to a head with the publication in English of Immanuel Kant's (1790, in German it was 1781) book *'The Critique of Pure Reason'* written in response to David Hume's (1757) philosophical scepticism. The point of disagreement was related to the idea that the mind itself gave structure to perceptions, providing concepts of time, space and causality, without which any experience would be chaotic and unintelligible. Kant (1790) indicated that scientific logic was relevant only to things people could experience through the senses and were able to prove/disprove by experiment. It failed to deal with value problems like beauty, *i.e.* reason could prove contradictory propositions and knowledge itself, as stated, was shaped by the senses and reason alone. Senses experience matter in space and time, which are subjective modes of sense

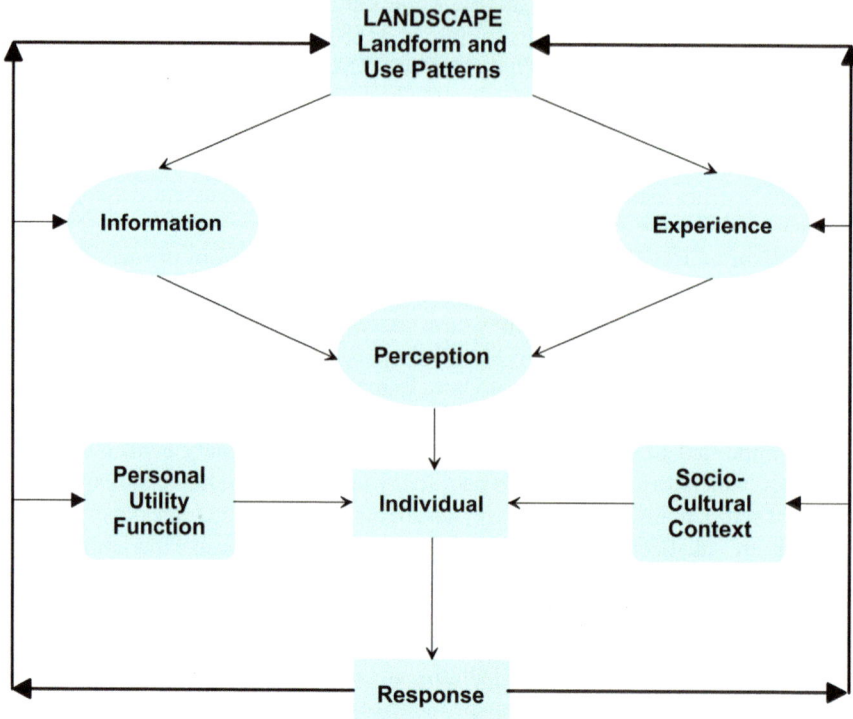

Fig. 2.4 Analysis of individual perceptions. (Based on Zube and Sell 1986)

perception and have no reality outside of the mind, as they are modes of perception and not attributes of the physical world. Reason also operates through logical structures, which also do not exist outside of the mind. This was a revolution in thinking, as fundamental as the revolution wrought by Copernicus in astronomy, *i.e.* it put nature and limitations of the mind at the centre of human knowledge. Wittgenstein (1922) in his *'Tractus Logico Philosophicus'*, carried Kant's (1790) work further by demonstrating the inadequacy of language itself to express the most important truths, arguing that language makes reasoning impossible on the fundamental issues, because the structure of syntax and imprecision of words cannot produce meaningful propositions about reality. Language does play a part, as words describing the trappings of landscape have to combine with those of feelings/emotion that are aroused in the beholder so terms, such as, topophilia came into usage (Tuan 1977).

Hume (1757) claimed that it was only force of habit that made us see a causal link behind all natural processes; man could not experience the law of causality. He attempted to *'reconcile the subjectivity of taste judgement and the experience of an effect with some objective features of the thing deemed beautiful'* (Eco 2016, 277). Kant (1790) turned this into an attribute of human reason *i.e.* the law is external and absolute simply because human reason perceives everything that happens as a matter of cause and effect. Both Kant and Hume agreed that one cannot know with

certainty what the world is like '*in itself*' – you only know what it is like '*for me*'.
Hume (1757) argued that man has two different types of perceptions, namely
impressions and ideas. The former relates to immediate sensations of an external
reality; the latter – the recollection of such impressions. External reality smacks of
phenomenology, a German philosophy propounded by Edmund Husserl and Martin
Heidegger (1987) focusing on the world as it appeared rather than questioning the
interpretations of reality – a philosophy that went straight to life and was the basis
of the post WWII existentialism movement, as espoused for example, by Sartre,
Camus and de Beauvoir.

George Berkeley, a sixteenth century Irish bishop, conforming to the ideas preva-
lent at the time, felt that science (and philosophy) were a threat to Christian life
(Atherton 1995). He questioned *via* the logic of empiricism, the viewpoint of people
like Descartes, Hume and Spinoza, who had argued that the material world is a reality. Berkeley stated that the only things that exist are those that we perceive and one
does not perceive '*material*' or '*matter*,' *i.e.* one can hit a table; it is hard, but you do
not feel the '*matter*'. He also questioned whether '*time*' and '*space*' had an absolute
or independent existence – they could be figments of the mind. However the key
figure was Descartes, a rationalist convinced that reason was the only path to knowledge. He was concerned with what we know, knowledge and the relationship
between body and mind. One could not accept anything as being true unless one
could clearly and distinctly perceive it. Every thought should be weighted and measured. Man was a dual creature, the mind was thought, and baser passions/feelings
e.g. desire, hate, beauty were considered extended reality. Remember Socrates' dictum, which related to one thing only – that he knew nothing.

Gray (1996) indicated that Isaiah Berlin (1909–1997) disagreed with the above
statements, as Berlin argued that concepts and categories by which one orders the
world do vary in time and space. Human conduct can only be explained by getting
inside a culture and its particular beliefs and values. Social science, searching for
general, abstract laws, was bound to miss what was peculiar and specific. Berlin
argued that everyone is faced by choices, between incompatible and incomprehen-
sible values and there is no single right answer to moral and political questions. If
you pursue community you sacrifice autonomy, indulge in mercy and justice goes,
so if one aspires to knowledge, one should be prepared to give up freedom.

In essence, with the publication of Kant's (1790) book, the viewpoint shifted to
'*the subject and the psychological state of the appreciator*', (Kant 1790, 7). Kant
(1790) defined delineated beauty into two categories: dependent beauty (anhdn-
gende Schonheit), which presupposes some concept of what the object should be,
and free beauty (freie Schonheit), which did not. When a person contemplates an
object and finds it beautiful, a harmony exists between his imagination and understanding, of which he is aware from the delight taken in the object. It is not cognitive
because it rests on feelings not on argument. Imagination grasps the object and is
not restricted to any definite concept, as it springs from the free play of cognitive
faculties. This was exemplified as, '*Beauty …. signifies nothing more than the relation of our judgment to an object*' (Descartes, in, Tatarkiewicz 1972, 172). The idea
of beauty without functionality involves the non-cognitive perception of aesthetics,

functionality being entrenched in our evolutionary past, later developed in the ideas of, e.g. Kaplan and Kaplan (1989a, b). It should be noted that aesthetic is time/culture dependant; the Greeks had their own aesthetic canons; the Romans a slightly different viewpoint; and in the Renaissance it differed again. Santayana (1896) rejected Kant's disinterested aesthetics, maintaining that the essence of aesthetics was 'pleasure objectified', experienced through the perception of an object, *i.e.* a value that only existed in perception.

But what in essence is 'Perception?' Oxford Dictionaries (2016) defines it, *'as the ability to see, hear, or become aware of something through the senses. In other words, it is the way in which something is regarded, understood, or interpreted, or the act or faculty of apprehending by means of the senses or of the mind. Another way to describe perception is as immediate or intuitive recognition or appreciation, as of moral, psychological, or aesthetic qualities; insight; intuition; cognition; understanding; realization, thoughtfulness.'* It is the act or faculty for perceiveng, refering of sensation to the external stimuli.

Therefore perception is not only a psychologial process but, also a cultural one. One sees *via* a personnal and cultural lens, according to lifetime habits which are shaped by a particular culture. Vision is the sense that both operates at a distance and close up unlike touch and taste and it can be focused and directed unlike hearing and smell. One can pay attention/ignore certain noises, smells, but one cannot blink ears or aim the nose; the eye is far more active and under far more control. Seeing is by far the best perceptual tool and the foremost way to engage with the world. It is an act of mind and not simply the eye, as information and instructions are travelling along nerves from the eye to the brain and *vice versa*. The fusion between object and observer, thought and emotion and action and reflection has been called synaesthesia (Lemley 1999; Girod et al. 2002) *i.e.* the blending of science and arts to increase awareness and perception and several pedagogies have highlighted the role of perception (Sullivan 2000).

It is salutary to remember the quote of Voltaire regarding individual perception of absolute beauty, the *'to kalon'*, *'Ask a toad what beauty is…He will answer that it is female, with two large round eyes sticking out of her little head and flat snout, a yellow belly, a brown back. Question the devil: he will tell you that the beautiful is a pair of horns, four claws, and a tail. Finally, consult the philosophers: their answer will be…grandiloquent nonsense; they ask for something conforming to the archetype of the beautiful, in essence, to the to kalon.'* (Voltaire 1824). Beauty and ugliness can therefore depend upon one's viewpoint and man posits himself as the norm of perfection (Nietzsche 2008), so beauty is defined with reference to a specific model.

Over the last two centuries the term perception has been utilised more and more in the literature. The term perception has been available since the 1800s, but post World War II an increase in its use may be linked to an increase presence of the term in web based media; it is likely related to journals/books with respect to scenery (Fig. 2.5; Ingold 2000). It is perception (positive or negative) that influences an individual or communities behavior irrespective of the scientific fact(s), which enables him/her to make a decision, *i.e.* what matters is what is believed, not what

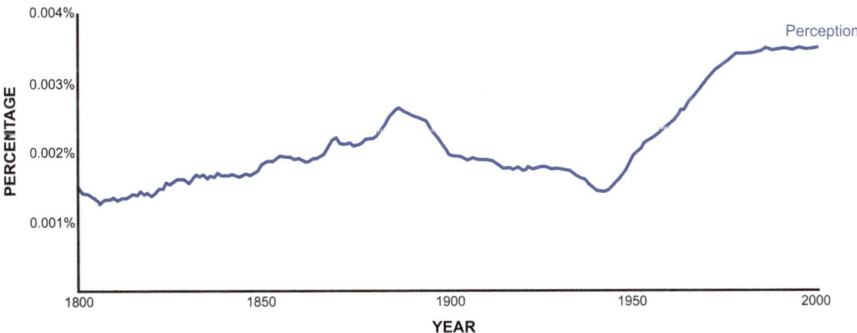

Fig. 2.5 Use of the term "perception" since 1800 (Google 2013)

is necessarily true, exemplified in the 2017 political world of the 'post truth' era where emotion has seemingly become more important than knowledge.

All that the above stresses is that it is the visual perception of a scene that is important and it is this that provides an impression of beauty, *i.e.* beauty has a special relationship with sight, a phrase associated with Descartes (Cottingham et al. 1991). Da Vinci (1891) argued that *'all our knowledge has its origins in our perceptions'*; Berger (1972, 7) reasoned that....*' the relationship between what we see and what we know is never settled'* adding *'that the 'way we see things is affected by what we know and what we believe'* (Berger 1972, 9) and a large part of seeing depends upon habit and convention. This infers that part of what we perceive comes through our senses from the object before us, another part comes out of our own head confirming Kant's (1790) affirmation that all judgments about beauty are individual judgments and rely on imagination. Is this *'the only faculty we have for apprehending beauty'* (Faulks 1999, 235)? The perception of beauty is influenced by experience. This is a key principle of the views of the seminal papers of Lowenthal (1961) and Appleton (1975), the latter adding an interdisciplinary approach not being deterred by seeking elementary information.

Kant (1790) saw in aesthetics an answer to a fundamental schism between pure and practical reason. So is it true that *'beauty properly denotes the perception of some mind'* (Stolnitz 1961, 201) and Margaret Wolfe Hungerford's – nee Hamilton (1889, 132) famed comment that *'Beauty is in the eye of the beholder?'* This has been an idea that has been expressed by many eminent people over a long time period:

- *'Beauty is bought by judgement of the eye/Not utter'd by base sale of chapmen's tongues'* (Shakespeare 1588).
- *'Beauty is no quality in things themselves: It exists merely in the mind which contemplates them; and each mind perceives a different beauty. One person may even perceive deformity, where another is sensible of beauty'* (Hume 1757, I. XX111.8).

- *'Beauty has been very often and very differently defined – and as often declared to be indefinable. If, however, we look for the characteristic defining relations, we find that the definitions hitherto suggested reduce conveniently to sixteen.'* (Ogden 2013, 216).
- *'Beauty …. signifies nothing more than the relation of our judgment to an object.'* (Descartes 1630).
- *'….whether things are beautiful because they please, or please because they are beautiful. And here I shall doubtless receive the reply that they please because they are beautiful.'* (St. Augustine).
- *'Beauty …. pertains rather to the sight than to the hearing.'* (Ficino 1561, quoted in Tatarkiewicz 1972).
- *'Everything has beauty, but not everyone sees it.'* (Confucius, www.brainyquote).
- *'I believe there is such a thing as a search for beauty – a delight in the nice things in world.'* (Leiter 2014).
- *'The most beautiful experience we can ever have is the mysterious. It is the fundamental emotion which stands at the cradle of true art and true science. Whoever does not know it can no longer marvel, is as good as dead, and his eyes are dimmed'.* (Einstein 1973, 12).
- *'Nec spe nec mytv',* (Plato 27, in Taylor 1929).
- *'Beauty in the landscape is a quality that appeals to the universal in man.'* (Younghusband 1920, 341).

So can beauty be quantified bearing in mind that the perception of beauty is subjective? *'Beauty is not an easy thing to measure. It does not show up in the gross national product, in the weekly pay check…. but these things are not ends in themselves. They are a road trip to satisfaction and pleasure and the good life. Beauty makes its own contribution to these final ends. Therefore it is one of the most important components of our true national income, not to be left out simply because statisticians cannot measure its worth'* (Johnson 1965, 10).

Does the word 'beauty' have a scientific definition and if so can it be measured and will its results/meaning that can allegedly be sensed, remain the same for the scientist as it would for other people? Understanding aesthetics in science requires consideration of sensual and emotional responses to nature similar, if not identical, to those involved in considerations of aesthetics in the arts (McAllister 1996), so to what extent is something beautiful or aesthetically pleasing? For Poincaré (1946) nature was beautiful, but was this nature in its entirety or minus the summation of anthropogenic effects? It appears contrary to common opinion that some of nature is beautiful and some is not. Can the use of the word 'beauty' to describe everything ultimately render the term meaningless (Kane 1997).

Plato (Taylor 1929) thought that mathematical beauty was the highest form of beauty, but on this score how many would find the following Eq. (2.2) beautiful.

$$(i\partial - m)\psi = 0 \qquad (2.2)$$

Yet this equation – The DIRAC equation – is aesthetic, elegant, simple and utterly beautiful to any mathematician, bringing together the two fulcrums of modern physics, quantum mechanics and relativity (ψ is the amplitude of the (four component) particle wave function; m is the (rest) mass of the particle; i is the square root of −1; the Feynman notation ∂ is the covariant derivative). What physicists/mathematicians love about the latter is its great mathematical beauty, which they have no difficulty in appreciating. What is important is the ability to identify simplicity in complexity, patterns in chaos, structure. Dirac commented that, *'what makes the theory of relativity so acceptable to physicists in spite of its going against the principle of simplicity is its great mathematical beauty. This is a quality which cannot be defined, any more than beauty in art can be defined, but which people who study mathematics usually have no difficulty in appreciating'* (BBC 2014). He argued that a physical law *must* possess mathematical beauty, *'It is more important to have beauty in one's equations than to have them fit the experiment;'* (Dirac 1963, 47). On that score he deemed the quantum electrodynamic (QED) theories of eminent scientists, such as, Heisenberg, Pauli, Feynman, Freeman Dyson (all Nobel Laureates) as 'ugly' and not worth doing anything with according to the Poincaré viewpoint.

Similarly the EULER equation (2.3) boasts five of the most important mathematical constants – zero (additive identity), 1 (multiplicative identity), e and pi (two of the most common transcendental numbers) and i (fundamental imaginary number), plus addition, multiplication and exponentiation; e, i, and π are extremely complicated and seemingly unrelated numbers, yet in this equation are inexorably linked together.

$$e^{i\pi} + 1 = 0 \qquad (2.3)$$

Greenbie (1975) suggested that neural patterns of emotion have a characteristic shape within the brain which can be recognised in similar shapes in the landscape and produce similar emotional responses. Some 30 years later Kawabata and Zeki (2004) used the technique of functional Magnetic Resonance Imaging (fMRI) to address the question of whether certain brain areas are specifically engaged when subjects view paintings classed as beautiful, neutral and ugly. fMRI exploits changes in bloodflow and oxygenation in order to identify regions of brain activity and represents a second generation in diagnostic medicine. They found that differential engagement existed for *'the medial orbito-frontal cortex, anterior cingulate, and the parietal cortex,'* so that *'the perception of stimuli as beautiful or ugly mobilized the motor cortex'* (Kawabata and Zeki 2004, 1700). Ten years later, Zeki et al. (2014) employed fMRI to show the activity of brain scans of 16 postgraduate or postdoctoral level mathematicians, as they looked at 60 formulae that previously had been judged as beautiful, half-way, or ugly. Results indicated that beautiful formulae stimulated activity in same brain field as experienced in the Arts.

In the Arts, 'beauty' can be accounted for, at least in part, by well-understood harmonies, distributions of colours or other factors. Blood et al. (1999) and Ishizu and Zeki (2011), showed that the same responses occur in music, Tsukiura and Cabeza (2011) for aesthetic and moral judgements, so these aspects of beauty all correlate with activity in a specific part of the emotional brain. This again was field

A1 of the medial orbito-frontal cortex (mOFC), which is the home of the experience of all beauty, *i.e.* a complex string of numbers/letters in mathematical formulae can provide the same sense of beauty as any art masterpiece or great music. There seems to be an abstract quality to beauty that is independent of culture and learning. Incidentally, Riemann's 'functional equation' was rated as the ugliest of the formulae tested! All aesthetic scientific responses seem in part to come from identifying simplicity in complexity, pattern in chaos, structure in stasis.

These objectively observable events turn into a first-hand feeling of a private subjective perspective experience and how this occurs is still unknown. Dearden (1985) rejected the '*objectivist*' and '*subjectivist*' philosophical poles preferring a '*relational*' one focused on the degree of societal consensus on landscape aesthetics, "*once thought of as ingrained, but are not so*" (Leopold letter 1975, pers.comm) concluding that familiarity was a significant factor in determining landscape preference. This tension between the objective and subjective has seemed insoluble since Descartes first wrote about it in the C17th when he segregated a conscious mind from a biological brain (Descartes in, Tatarkiewicz 1972). Perception is about what is seen or experienced. A brain can be reduced to neurons sparking/pulsating with chemicals but somehow this seething neurobiological mass relates to consciousness that is experienced, e.g. in a coastal scene, with the sound of waves breaking on a shore, the sight of a soaring vertical coastal cliff – a conceptual impasse indeed (Greenfield 2016).

In 1970, Colvin (1970, xxii) wrote that '*the study of landscape design must now bridge the chasm between art and science, it is even more necessary for science and art to come to terms.*' It appears that describing something as beautiful – a certain landscape, painting, formula or music perhaps – tends to promote the same pleasure in an observer's brain, so the chasm of Colvin (1970) appears to have been bridged.

One must remember that any landscape view '*is composed not only of what lies before our eyes but what lies within our heads*' (Meinig 1979, 58) and different people e.g. a geologist, ecologist, archaeologist, will see landscapes in different ways *via* nature, habitat, wealth, artefact, history, etc. This is geared to intuition and experimental thinking guided, as stated, by emotional and affective processes in the brain, which is an emotional entity, a vast array of neurons that work together to generate our experience of the world. Associations of networks, bundles of thoughts, feelings, images and ideas become connected over time. Therefore the brain, '*is the bridge between the internal and external universes,*' (Martin and Gordon 2001, 393) covering three scales: MACROSCOPIC- Cosmic; MESOSCALE- Global; MICROSCOPIC- Local.

External (Fig. 2.6)

(a) Intensity and Size. The brighter a light, the more likely one is to see it.
(b) Contrast and novelty. New Stimuli will often gain one's attention, as can the appearance or disappearance of stimuli.
(c) Repetition. Repeated stimuli can cause one to/pay attention, e.g. the blue flashing light of an ambulance or police car.

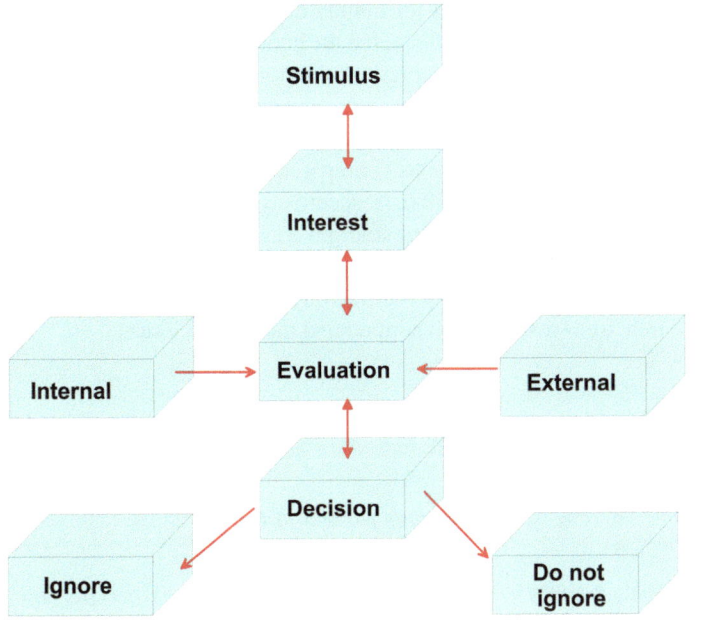

Fig. 2.6 The perception process

Internal (Fig. 2.6)

(a) Notices and needs e.g. hunger, thirst.
(b) Preparatory set. A person can awake on to early morning call if it is expected
(c) Interest. Attention is paid to football scores, whilst effectively ignoring the rest of a radio/tv programme.

Zube et al. (1982) reviewed 160 journal papers from UK and North America, with reference to contributions to pragmatic landscape planning and management issues and to the evolution of a general theory of landscape perception and identified four general paradigms of landscape perception.

- Expert evaluation by skilled observers, e.g. Taggert et al. (1980), Shuttleworth (1980a, b).
- Psychological, assessment *via* testing the public with respect to aesthetic landscape qualities, e.g. Carls (1974).
- Cognitive pairing involving the search for a human meaning associated with landscape properties, e.g. Wohill and Harris (1980).
- Experiential which considered landscape values to be based on experience of the human-landscape interaction where both were shaping and being shaped in the interactive process, e.g. Kobayashi (1980).

Zube et al. (1982) proposed a theoretical framework to guide future research to looking at how human-landscape-outcome actions interrelate, concluding that the justification for worrying about landscape perception and making landscape beautiful is that landscape is important for human quality of life and is as significant as economic and social factors in influencing the human condition. They suggested that because there is no underlying theoretical structure for landscape perception, there is a lack of a rational basis for diagnosis, prescription and prognosis. They even posed the question as to whether researchers were measuring the same aesthetic.

From 1965–80, Zube et al. (1982) compared the ratio of the paradigms for the 1965–'73 periods with the 1974–'80, (respectively they were: 3:1; 8:1; 4:1; 5:1) and results suggested that the dominant paradigm was the psychological and the expert approach was popular. This reflected an emphasis on problem solving research and Landscape journals have primarily focused on expert and psychological paradigms. These research areas involve both applied and theoretical issues.

2.2.1   The Legal Situation

Fresh from brawling courts/ and dusty purlieus of the law. Alfred Lord Tennyson, In Memoriam

This varies from country to country and the situation for the UK is presented simply as an example. The first formal UK legislation Act where the term 'natural beauty' was the 1906–7 Act to establish the National Trust, a charitable organisation for places of historic interest and natural beauty, whose philosophy was based upon the ideas of Robert Hunter and Octavia Hill. Since then many committees and landmark Acts have been formulated, e.g. Government Reviews e.g. Select Committee Public Works, 1833; Addison Committee, 1929; Scott Committee, 1941, Dower Report, 1945 (Williams 1987). Terms such as *'characteristic landscape beauty is strictly preserved,' 'landscape value and rural areas of remarkable landscape beauty;' 'landscape character and landscape pattern;' 'high landscape quality;' 'high scenic value,'* were bandied about in these reports with respect to natural beauty, all leading to the UK 'Town and Country Planning Act, (1947)'. This was followed by the Hobhouse Report, 1947, and the major National Parks and Access to the Countryside Act (1949), a fundament for UK legislation whose implications can be seen in the IUCN (1994) 'protected areas 'of the Category V protected landscapes/seascapes that are seemingly more anthropogenic than natural.

In 1949, the Town and Country Planning Act, received Royal Assent and Section 5(1) stated that the National Parks' purpose included, *'the preservation of the natural beauty of an area'*, qualified in Section 114(2) by, *'references in this act to the preservation of the natural beauty of an area shall be construed as including references to the preservation of the characteristic natural features, flora and fauna thereof,'* amended in 1968 to *'its flora, fauna and geological and physiographical features'* in which 'preservation' was replaced by 'conservation'. Interpretation of Section 114(2) is that it is a limited definition of 'natural beauty' and does not flesh out the meaning but the phrase cropped up in many later Acts, notably the Countryside Act (1968), Environment Act (1995), Wildlife and Countryside Act

(1981 and 1985), Countryside and Rights of Way Act (2000), together with accompanying EC regulation relating to Environmentally Sensitive Areas.

The meaning has inevitably been assumed to be self-evident i.e. beauty that is natural and not man made, but within the past century its legal definition has been challenged and despite widespread legislation usage it has not been formally defined. The crux of the matter has been in deciding at what point does a landscape cease to be 'natural' because of 'human settlement and land use'. In a report concerning an inquiry into the North Pennines Areas of Outstanding Natural Beauty, Berthon (1986) argued that, *'the quality of the landscape of the area is both natural and man-made'*, i.e. quality in place of natural beauty and man-made made an appearance. A second review of National Parks (Edwards 1991) recommended that the first purpose of National Parks, of *'preserving and enhancing the natural beauty of the areas'* should be re-defined as *'to protect, maintain and enhance the scenic beauty, natural systems and landforms, and the wildlife and cultural heritage of the area.'* So culture now becomes very important. 'Natural beauty' was qualified by Section 114(2) indicating that in all references in the Act to the preservation/conservation of the 'natural beauty' of an area, 'natural beauty' would include, but was not confined to, flora and fauna and geological and physiographical features.

'Landscape' and 'character' now commenced to be seen as being related but having different terms, and the problem of the application of worldwide aesthetic rules became evident. Beauty related to quality whilst character was concerned with individual landscape qualities and consequent values. In an appeal decision in a case relating to The New Forest boundary, the High Court (2005) suggested that legislation was needed to clarify the meaning and use of natural beauty in designating landscapes for protection. It stated that *'views as to which tracts of countryside have the quality of 'natural beauty'* may (or may not) have changed over the last 50 years, but the 'natural beauty' criterion in subsection 5(2)(a) of the Act has not been changed to embrace wider considerations such as 'cultural heritage'. If the 'natural beauty' criterion in subsection 5(2)(a) is to be changed to reflect twenty-first century approaches to countryside and leisure planning, then the change must be effected by Parliament, and not by administrative action on the part of the Agency in adopting a wider range of factors for the purposes of designation.' Recently the words 'natural beauty' in a superb paper by Selman and Swanwick (2010, 4) referred to a *'dynamic and malleable concept potentially posing problems for consistency of interpretation and yet apparently retaining a continuing relevance'*. This is currently the UK situation.

2.3 Conclusions

The phrase that is guaranteed to wake up an audience…And in Conclusion, (Anon)

Many branches of the arts/sciences have debated the term 'beauty', e.g. philosophers, artists, scientists, whilst geographers have attempted to apply it to landscapes. The concept of beauty is difficult to resolve as it relates to an individual's perception

Fig. 2.7 (**a**) A 'beautiful' scene, Anakena beach, Easter Island, Chile; (**b**) An 'ugly' scene, Ventanas, Valparaiso Region, Chile

of a scene, which encompasses the individual components of a scene that interacts with the brain of the person by relating it to past experiences. This is the reason why field testing of the views of many beach users was used in this book – see Chap. 4. An agreement of opinion from this cohort – all geared to their personal perception of coastal scenery allowed a consensus as to what constitutes a beautiful coastal scene and its converse an ugly scene. Most people if presented with Fig. 2.7a intuitively would rate as a scene exhibiting 'beauty', whereas Fig. 2.7b would be described as 'ugly.'

Any coastal landscape will encompass physical, historical and cultural parameters and many of these parameters are ephemeral, e.g. the smell of salt air, the slow passage over water of a white sailed yacht – these are impossible to quantify, but the visual impact of a coastal landscape remains the most dominant of the senses.

However, it is prudent to remember the words penned by Maugham (1963, 96–97) … *'the ideal has many names and beauty is but one of them…I cannot contemplate beauty long.…when the thing of beauty has given me the magic of its sensation my mind quickly wanders.… Beauty is an ecstasy; it is as simple as hunger.… Beauty is perfect, and perfection (such is nature) holds our attention but for a little while, beauty is that which satisfies the aesthetic instinct .… beauty is a bit of a bore.'* Perhaps the meaning of the word is not to be found in some abstract definition but in its usage in everyday language, something akin to Wittgenstein's (1922) viewpoint, remembering that everyday speech has embedded biases., but then, … *que sais-je*? (de Montaigne 2016).

Addendum

This chapter really only refers to western ideas of beauty. Ideas from e.g. the eastern world, are unfortunately not encompassed in this discussion mainly due to the author's inability for language.Appreciation of, for example, an Asian/Polynesian/Maori/Aborigine concept of beauty is especially true with international tourism expected to reach 1.8 bn. by 2030 (UNWTO 2017), with an expected huge influx from Asian countries.

References

Amir S, Gidalizon E (1990) Expert based method for the evaluation of visual absorption capacity of the landscape. J Environ Manag 30:251–163

Appleton J (1975) Landscape evaluation: the theoretical vacuum. Trans Inst Br Geogr 66:120–123

Appleton J (1996) The experience of landscape. Wiley, London. 296pp

Atherton M (1995) Berkeley without God. In: Muehlmann G (ed) Berkeley's metaphysics: structural, interpretive, and critical essays. Pennsylvania State University Press, Philadelphia, pp 231–248

Barnes J (2015) Keeping an eye open. Jonathon Cape, London. 288pp

Baumgarten AG (1735) Theory of aesthetics, philosophical study of art and natural beauty. 1954 edition: (trans: Aschenbrenner K, Holther WB). University of California Press, Berkeley

BBC (2014) Mathematics: why the brain sees maths as beauty. BBC, London

Berger J (1972) Ways of seeing. Penguin, London. 176pp

Berthon Sir S (1986) North Pennines AONB (Designation) Order 1978: report of Public Local Inquiry (Reference: DRA1/CW/39)

Blackmore RD (1993) Lorna Doone: a romance of Exmoor. Wordsworth Classics, London. 247pp

Blood AJ, Zatorre RJ, Bermudez P, Evans AC (1999) Emotional responses to pleasant and unpleasant music correlate with activity in paralimbic brain regions. Nat Neurosci 2:382–387

Böhme G (2010) On beauty. Nordic J Aesthet 21(39):22–33

Boulton JT (1958) Edmund Burke: a philosophical enquiry into the origins of our ideas on the sublime and the beautiful. Routledge and Kegan Paul, London. 197pp

Bufford S (1973) Beyond the eye of the beholder: aesthetics and objectivity. Mich Law Rev 71:1438–1463

Buhyoff GJ, Arndt LK (1981) Interval scaling of landscape preference by direct and indirect measurement methods. Landsc Plan 8:257–267

Buhyoff GJ, Hull RB, Lien JN, Cordell HK (1986) Prediction of scenic quality for southern pine stands. For Sci 32(3):769–778

Burke E (1757/2008) A philosophical enquiry into the origin of our ideas of the sublime and beautiful. Oxford University Press, Oxford

Carls EG (1974) The effects of people and man-induced conditions on preferences for outdoor recreational landscapes. J Leis Res 6:113–124

Cassells dictionary (1997) Orion, UK

CC – Countryside Commission (1987) Landscape assessment 2013 a countryside commission approach. Countryside Commission, London

Chambers dictionary (2000) Chambers, UK

CoE – Council of Europe (2000) European landscape convention. Council of Europe, Florence

Colvin B (1970) Land and landscape. John Murray, London. 266pp

Cosgrove D, Daniels S (1988) Introduction: iconography and landscape. In: Cosgrove D, Daniels S (eds) The iconography of landscape. Cambridge University Press, Cambridge, pp 1–10

Cottingham J, Stoothoff R, Murdoch D (1991) The philosophical writings of Descartes, vol 3. Cambridge University Press, Cambridge

Cox G (1988) Reading nature: reflections on ideological persistence and the politics of the countryside. Landsc Res 13(3):24–34

Da Vinci L (1891) Codex Trivulzi, Castello Sforzesco. Pub. Luca Beltrami, Milan

Davy H (1840) Parallels between art and science. In: Davy J (ed) The collected works of Sir Humphrey Davy, vol 8. Smith and Cornhill, London, pp 306–308

de Montaigne M (2016) The complete essays by Michel de Montaigne. Create Space Independent Publishing Platform, London

Dearden P (1985) Public participation and scenic quality analysis. Landsc Plan 8:3–19

Descartes letter by R. Descartes to M. Mersenne, in, Tatarkiewicz (1972)

Descartes R (1630) Letter to Mersenne

Dewey J (1929) Experience with nature. Allen and Unwin, London. 486pp

Dirac PAM (1963) The evolution of the physicist's picture of nature. Sci Am 208(5):45–53

Duncan JS, Duncan NG (1992) Ideology and bliss: Roland Barthes and the secret histories of landscape. In: Barnes TJ, Duncan JS (eds) Writing world: discourse, text and metaphore in the representation of landscape. Routledge, London, pp 250–266

Eckbo G (1975) Qualitative values in the landscape. In: Zube EH, Brush RO, Fabos JG (eds) Landscape assessment. Dowden Hutchinson and Ross, Stroudsburg, pp 151–167

Eco U (2011) On ugliness. Rizzoli, New York. 453pp

Eco U (2016) History of beauty. Rizzoli, New York. 438pp

Edwards RC (1991) Fit for the future: report of the National Parks Review Panel. Countryside Commission, Cheltenham

Einstein A (1973) Ideas and opinions. Souvenir Press, London

Eisner T (1999) Seventy-five reasons to become a scientist. Am Sci 76:451

Ergin A, Williams AT, Micallef A (2006) Coastal scenery: appreciation and evaluation. J Coast Res 22(2):958–964

Fairbrother N (1970) New lives, new landscapes. Architectural Press, London. 397pp

Faulks S (1999) Charlotte Grey. Hutchinson, London. 393pp

Ficino M (1561) quoted in Tatarkiewicz 1972

Foster CA (1991) Aesthetics and the natural environment. Unpublished PhD. thesis, University of Edinburgh

Girod M, Rau C, Schepige A (2002) Appreciating the beauty of scientific ideas: teaching for aesthetic understanding. Sci Educ 87:574–587

Google (2013) https://books.google.com/ngrams. Accessed 16 May 2017

Gray J (1996) Isaiah Berlin. Princeton University Press, Princeton. 183pp

Greenbie BB (1975) Problems of scale and context in assessing a generalized landscape for particular persons. In: Zube EH, Brush RO, Fabos JG (eds) Landscape assessment. Dowden Hutchinson and Ross, Stroudsburg, pp 168–201

Greenfield S (2016) A day in the life of a brain. Allen Lane, London. 288pp

Guyer P (2004) The origins of modern aesthetics. In: Kivey P (ed) The Blackwell guide to aesthetics. Blackwell, London, pp 1711–1735

Hartshorne R (1939) The nature of geography. Assoc Am Geogr Lanc USA 29(3):173–412

High Court (2005) High Court judgment in Meyrick Estate Management & Others v. Secretary of State for Environment, Food and Rural Affairs [2005] EWHC 2618 (Admin), Paragraph 62

Hogarth W (1753) The analysis of beauty. J. Reeve, London

Socrates. http://www.academiasocrates.com/socrates/biografia.php

Confucious. http://www.brainyquote.com/quotes/authors/c/confucius

Hull RB, Revell GRB (1989) Issues in sampling landscapes for visual quality assessments. Landsc Urban Plan 17:323–330

Hume D (1757) Of the standard of taste, reprinted in A. Neill, and A. Ridley (1995). The Philosophy of Art: Readings Ancient and Modern. McGraw Hill, Boston, 592pp

Hungerford MW (1889) Molly Bawn (Dodo Press). Hurst and Blacket, London

Husserl E, Heidegger M (1987) The phenomenology of internal time-consciousness. University Microfilms International, Ann Arbor

Ingold T (1993) The temporality of the landscape. World Archaeol 25:152–174

Ingold T (2000) The perception of the environment: essays of livelihood, dwelling and skill. Routledge, London. 460pp

Ishizu T, Zeki S (2011) Toward a brain-based theory of beauty. PLoSONE 6:e21852

IUCN (1994) International Union for the Conservation of Nature and Natural Resources (IUCN). Guidelines for protected area management categories. IUCN, Gland

Johnson LB (1965) Special message to the congress on conservation and restoration of natural beauty. February 8. http://www.presidency.ucsb.edu/ws/?pid=27285

Jovaisa M (2015) Unseen Cuba. Unseenpictures, Berlin. 440pp

Kane M (1997) Beauty and science. Philosophy Now 17:15–19

Kant I (1790) Critique of Judgment (Kritik der Urteilskraft), 1st edn. L. Lagarde, Berlin, p 480

Kaplan R, Kaplan S (1989a) The visual environment: public participation in design and planning. J Soc Issues 45(1):59–86

Kaplan R, Kaplan S (1989b) The experience of nature: a psychological perspective. Cambridge University Press, Cambridge. 360pp

Kates RW (1966) The pursuit of beauty in the environment. Landscape 16(2):21–25

Kawabata H, Zeki S (2004) Neural correlates of beauty. J Neurophysiol 91(4):1699–1705

Kaye R, Alder J (1999) Coastal planning and management. E & FN Spon, London. 375pp

Keats J (1918) Endymion, London, Book I, lines 1 and 2

Kobayashi A (1980) Landscape and the poetic act: the role of the haiku club for the Issei. Landscape 24:42–47

Langer SK (1953) Feeling and form: a theory of art developed from philosophy in a new key. Routledge and Kegan Paul, London. 431pp

Leiter S (2014) No Great Hurry. http://watch.innogreathurry.com

Lemley B (1999) Do you see what they see? Discover 20(12):80–87

Leopold LB (1969) Quantitative comparisons of some aesthetic factors among rivers. US Geological Survey, Washington, DC. 16pp

Levi P, Regge T (1989) Dialogo (trans: Rosenthal R). Princeton University Press, Princeton

Linton DL (1968) The assessment of scenery as a natural resource. Scott Geogr Mag 84:219–238

Linton DL (1982) Visual assessments of natural landscapes. W Geog Ser 20:97–116

Locke J (1816) An essay concerning human understanding. Thomas Tegg, London. 816pp

Lodge D (1992) The art of fiction. Penguin, London. 239pp

Lothian A (1999) Landscape and the philosophy of aesthetics: is landscape quality inherent in the landscape or in the eye of the beholder? Landsc Urban Plan 44:177–198

Lowenthal D (1961) Geography, experience, and imagination: towards a geological epistemology. Ann Assoc Am Geogr 51:241–260

Lowenthal D (1972) Geography, experience and imagination: towards a geographical epistemology. In: Ward English P, Mayfield RC (eds) Man, space and environment. Annals of the American Association of Geographers, Washington, DC, pp 219–244

Martin CC, Gordon R (2001) The evolution of perception. Cybern Syst Anal 32:393–409

Maslow A (1968) Towards a psychology of being. Van Norstrand, New York

Maugham S (1963) Cakes and Ales. Penguin, London

McAllister JW (1996) Beauty and revolution in science. Cornell University Press, Ithaca. 231pp

McEvoy J (2000) Robert Grosseteste (Great Medieval Thinkers). Oxford University Press, New York

McGann J (2017) Poor human olfaction is a 19th-century myth. Science. http://science.sciencemag.org/. Accessed 18 May 2017)

Meinig RW (1979) The beholding eye: ten versions of the same scene. In: Meinig DW (ed) The interpretation of ordinary landscapes, geographical essays. Oxford University Press, Oxford, pp 33–48

Morisawa M (1971) Quantitative geomorphology: some aspects and applications. McGraw Hill, New York

Mowaljarlai DB (1995) An address to the white people of Australia', ABC radio: the Law report, Melbourne

Newton E (1966) The meaning of beauty. Whittlesly House, New York

Nietzsche F (2008) Twilight of the Idols. Duncan Large (ed). Oxford World's Classics, 176pp

Ogden CK (2013) Bentham's theory of fiction. Routledge, New York

Olwig K (2003) The landscape legacy. Ann Assoc Am Geogr 91(4):871–877

Oxford Dictionaries (2016) http://www.oxforddictionaries.com/definition/english-thesaurus/perception. Downloaded: March 2016. Oxford University Press

Penning-Rowsell EC, Lowenthal D (1986) Landscape meanings and values. Harper Collins Publishers, London

Poincaré H (1946) The Foundations of Science (trans: Halsted G). Science Press, Lancaster, PA

Porteous JD (1985) Smellscape. Prog Hum Geogr 9(3):356–378

Porteous JD, Mastin JF (1985) Soundscape. J Archit Plann Res 2(3):169–182

Purcell AT (1992) Abstract and specific physical attributes and the experience of landscape. J Environ Manag 34:159–177

Purcell AT, Lamb RJ (1998) Preferences and naturalness: ecological approach. Landsc Urban Plan 42(1):57–66

Santayana G (1896/1955) The sense of beauty. Being the outlines of aesthetic theory. Dover Publications, New York

Sauer C (1938) The morphology of landscape. University California Press, Berkeley

Selman P, Swanwick C (2010) On the meaning of natural beauty in landscape legislation. Landsc Res 35(1):3–26

Shakespeare WS (1588) Love's Labour's Lost, Act 11, scene 1

Shuttleworth S (1980a) The evaluation of landscape quality. Landsc Res 5:14–20

Shuttleworth S (1980b) The use of photographs as an environmental presentation medium in landscape studies. J Environ Manag 11:61–76

Shuttleworth S (1983) Upland landscapes and the landscape image. Landsc Res 8(3):7–14

Skylar M (1967) Walt Disney World: Background and Philosophy. Harrison "Buzz" Price Papers. Paper 160. http://stars.library.ucf.edu/buzzprice/160

Stern P (2017) Humans have a good sense of smell. Science 356:594–596

Stolnitz J (1961) 'Beauty': some stages in the history of an idea. J Hist Ideas 22(2):185–204

Sullivan AM (2000) Notes from a marine biologist's daughter: on the art and science of attention. Voices insides schools, Havard. Educ Rev 70(2):211–227

Taggert C, Tetherow T, Bottomley B (1980) Visual values: Colorado lakes stock. Landsc Archit 70:396–400

Tatarkiewicz W (1972) The great theory of beauty and its decline. J Aesthet Art Critic 31(2):165–180

Taylor AE (trans) (1929) Plato, Timaeus and Critias, 30b, London

Tsukiura T, Cabeza R (2011) Shared brain activity for aesthetic and moral judgments: implications for the Beauty is Goodstereotype. Soc Cogn Affect Neurosci 6:138–148

Tuan YF (1977) Space and place: the perspectives of experience. University of Minnesota Press, Minneapolis

Tudor C (2014) An approach to landscape character and assessment. Natural England, London

UNWTO (2017) (United Nations World Tourism Organization). Tourism highlights edition, 16 pp

Vandenabeele B (2003) Schopenhauer, Nietzsche, and the aesthetically sublime. J Aesthet Educ 37(1):90–106

Voltaire F (1824) Philosophical dictionary, beau, beauté; beauty, beautiful Vol 2. John and Henry Hunt, London

Waterfield R (trans) (1993) Plato Republic – translated with notes and an introduction. Oxford University Press, Oxford

Weinberg S (1993) Dreams of a final theory: the search for the fundamental laws of nature: search for the ultimate laws of nature. Vintage Press, London. 272pp

Williams AT (1987) Coastal conservation policy development in England and Wales with special mention of the heritage coast concept. J Coast Res 31(1):99–106

Williams AT, Micallef A (2009) Beach management: principles and practices. Earthscan, London

Willstaetter R (1965) From My Life. The Memoirs of Richard Willstaetter (trans: Hornig S, Benjamin WA) New York (originally published in German, Verlag Chemie, 1949)

Winton T (2016) Island home. Penguin Books, Melbourne

Wittgenstein L (1922) Tractatus Logico-Philosophicus. Kegan Paul, Trench, Trubner, New York

Wohill JF, Harris G (1980) Responce to congruity or contrast for mad –made features in natural recreation settings. Leis Sci 3:349–365

Wright S (2005) Reflections on the rock. Aust Geogr 80:61–86

Younghusband F (1920) Natural beauty and geographical science. Prog Hum Geogr 2(2):338–348

Zeki S, Romaya JP, Dionigi M, Benincasa MT, Atiya MF (2014) The experience of mathematical beauty and its neural correlates. Front Hum Neurosci B 68:1–12

Zube EH, Sell JL (1986) Human dimensions of environmental change. J Plan Lit 1(2):162–117

Zube EH, Sell JL, Taylor JG (1982) Landscape perception: research, applications and theory. Landscape Plann 9:1–33

Web Pages:

www.bbc.co.uk/news/science-environment-26151062. Accessed 30 July 2017

www.official-documents.gov.uk/. Accessed 30 April 2016

www.science.sciencemag.org/content/356/6338/caam7263. Accessed 17 May 2017

Chapter 3
Some Scenic Evaluation Techniques

Allan T. Williams

Mountains are the beginning and the end of all natural scenery.

John Ruskin, Modern Painters, vol.3, Pt iv, v 29.

Abstract The quest for an objective analysis of coastal scenery ranging from top class to very poor, which includes the physical environment and incites an aesthetic response from the viewer, has existed for many years and a variety of approaches have been employed. These range from utilizing photographs as surrogates for locations, to compiling lists of what are deemed to be the important physical and anthropogenic landscape parameters, either from field studies, Ordnance Survey map squares, or *via* questionnaires of public attitudes and perception, etc. The bulk of these studies have been aimed at producing numbers that can be attributed to relevant assessment parameters in an attempt to quantify landscapes. These include physical items, such as, relief and slope, which relate to mountains, hills, lowlands, as well as human induced elements, such as, towns, industrialized areas, and farmed landscapes. The spectrum covered by these techniques tends to be based on subjective assessments of landscape quality by individuals/groups, or by techniques using landscape physical attributes as surrogates for personal perception.

3.1   Introduction

Every day speaks a new scene; the last act crowns the play. Francis Quarles, Epigram Respice Finem.

Arthur et al. (1977) argued that any scenic assessment must have:

A. T. Williams (✉)
Faculty of Architecture, Computing and Engineering, University of Wales, Trinity Saint David, Swansea, Wales, UK

CICA NOVA, Nova Universidad de Lisboa, Lisbon, Portugal

© Springer International Publishing AG, part of Springer Nature 2019   43
N. Rangel-Buitrago (ed.), *Coastal Scenery*, Coastal Research Library 26,
https://doi.org/10.1007/978-3-319-78878-4_3

- Descriptive inventories that include ecological and formal aesthetic models, together with methods mostly applied by experts in an objective manner.
- Public preference models, such as, psychological and phenomenological, which are often undertaken using questionnaires are unavoidably linked to the problems of public consensus.
- Quantitative holistic techniques using a mixture of subjective and objective methods including psychophysical and surrogate component models.

All assessments should heed these words penned some 40 years ago. However, most scenic assessments have been carried out on a subjective basis; have no weightings attached to parameters (the latter invariably had the same value – wrong, as some parameters are more important than others); and none concerned itself with a technique that utilises fuzzy logic systems (Chap. 4), which look into the 'grey areas' rather than e.g. between black/white; true/false.

Landscape assessments procedures have usually been seen as a contrast between expert based opinions, usually based on ecology, which normally is the choice of management schemes and a perception based viewpoint, based on philosophical aesthetics. Non-expert judgements usually rely on methods based on landscape stimulus and the objective properties of landscape. Greenbie (1975) suggested that neural patterns of emotion have a characteristic shape within the brain which can be recognised in similar shapes in the landscape and produce similar emotional responses – see Chap. 2 and also Starr (2013). It is difficult to separate landscape experience from the viewpoint context and the emotional involvement, which makes for difficulty in developing techniques for methodological development other than unstructured phenomenological exploration. Human-landscape interaction suggests that aesthetic quality can occur in both the objective qualities of landscape and the subjective meaning of landscape, as observed many years ago by, for example, Lewis et al. (1973), Relph (1976) and Tuan (1977).

In the UK, the Countryside Act (1968, A2) stated: '*The functions conferred by this Act on the said Commission in this Act (referred to as "the Commission") are to be exercised for the conservation and enhancement of the natural beauty and amenity of the countryside, and encouraging the provision and improvement, for persons resorting to the countryside, of facilities for the enjoyment of the countryside and of open-air recreation in the countryside.*' This Act was a springboard to many other Acts, which resulted in the UK of a founding based on beauty, of e.g. Areas of Outstanding Natural Beauty (AONB) and Heritage Coasts (Fig. 3.1).

3.2   Background

Scenery is fine but human nature is finer. John Keats, letter to Benjamin Bailey, 13 March 1818

It is axiomatic that any landscape assessment invokes interaction between biophysical parameters and the perception that a viewer receives from looking at it. Both

Fig. 3.1 Cliffs and a shore platform at Glamorgan Heritage Coast, UK

involve the effects of spatial and temporal scales, which affect landscapes, especially the visual component. A psychophysical approach provides balance between the two viewpoints expressed, as advocated by the checklist given in Chap. 4, producing an equalising effect between the prime parameters needed for coastal landscape assessment, together with an objective account of a viewer's reaction. It is important to stress that there is nothing wrong in any person's subjective evaluation of any given landscape; however for management purposes the opinions of many people are required, as viewpoints from many individuals tend to conform to the required real and objective consideration.

Steers (1944, 6), after an 18 month journey around the coast of England and Wales emphasised that, '*any unspoiled part of the coast; cliffs, dunes, salt marsh, estuary, should be rated as good natural scenery. On the other hand certain parts of the coast can be regarded as of outstanding quality.*' He also added that, '*any assessment of coastal quality is likely to meet with criticism.*' This is certainly a truism, as Teale (1966, 72) also argued that, '*nature affects our minds as light affects a photographic emulsion on a film. Some films are more sensitive than others, some minds are more receptive.*' Davidson and Wibberley (1977, 87) followed up these comments suggesting that some landscapes appear intrinsically more valuable than others and this has been the basis of landscape conservation in many countries e.g. Britain. Therefore scenic beauty has become a very important factor relating to management decisions, in amongst others, National Parks (Fig. 3.2), Heritage Coasts, and has led to a proliferation of methodologies to predict landscape quality.

How is scenery evaluated, as it covers a spectrum of 'beauty/ugly', and how is this elusive concept called beauty defined – see Chap. 2; Fig. 3.3? It is far easier to agree on what constitutes ugliness, yet even scenes, such as, old coal tips, can on occasions, e.g. when covered by snow, take on an ethereal quality and even be con-

Fig. 3.2 Mwnt, Pembrokeshire National Park. Wales, UK

Fig. 3.3 Annaly Point, US Virgin Islands – a beautiful scene!

sidered *'beautiful'*, remembering Margaret Hungerford's (1889, 132) famous phrase, *'beauty lies in the eye of the beholder.'*

However, coastal/beach managers/planners have to take cognisance of this point with respect to any management affecting the world's coastlines. Decisions are taken with regard to coastal areas, but how much attention is paid to scenery; is it an important constituent for a pleasurable holiday – an emphatic 'Yes'; or if the decision is to zone an area as one of 'Outstanding Natural Beauty', 'Heritage Coast', 'National Park' etc., what criteria are utilised?

An early pioneer in this field was Hartshorne (1939) who looked at the separation between Natural and Human landscapes and Appleton (1975b, 2) who was one of the later followers commenting that *'Landscape is a kind of buckcloth of human activities,'* and his classic book, posed the question, *'what do we like about landscapes and why do we like it?'* Appleton (1975b, viiii). The history of landscape quality assessment/evaluation, *'has featured a contest between expert and perception based approaches, paralleling a long standing debate in the philosophy of aesthetics'* (Daniel 2001, 267). A first rate review of landscape methodologies can be found at Macaulay (2014), whilst excellent up to date analysis of landscape character assessment can be found in Swanwick (2002), Olwig (2003), and Tudor (2014). *'Inspired by the open beauty of the open landscape'* (Faulks 1999, 267), is a frequent literary comment, but how does one measure this?

Eckbo (1969, 19) mentions that *'...the tremendous volume of travel literature does demonstrate that the general quality of landscape can be measured and is measured, and that there is a fair unanimity as to their findings.'* The need for evaluation only arises when: *it becomes necessary to make a choice between two or more articles... or courses of action'* (Gillespie 1970, 25).

Evaluation is necessary, as it provides a means by which scenery/amenities can be compared against other resource considerations; it can improve resource inventories, carrying capacity decision-making and Environmental Impact Assessments (IEMA 2011). It is a process whereby landscape is weighed against set criteria, i.e. it is a statement of overall aesthetic quality of an object in comparison with other objects of the same class. It provides landscape preservation (identifying the value to society of particular views/areas); protection (identifying high quality landscapes/controlling developments); improvements (identification of components that can detract from views). Scenery is a section of any coastal landscape inventory available for managers or planners for coastal preservation, protection, development, etc. and can provide baseline information to managers so that a sound scientific basis may be established for any subsequently envisaged management plans. Landscape management and evaluation deals with heterogeneity in space, for example *via* type, shape, etc. of elements, together with time, e.g. the disturbance regime, which may be either natural or anthropogenic (Turner 1987).

A primary influence behind public concern for landscape quality and evaluation is an increasing rejection of older attitudes of man's dominance over nature in favour of a more mutual respect. This had been mirrored in an increasing environmental awareness, which reached a watershed in the 1970s when extensive environmental lobbying had a significant effect on public policy (Sewell and Foster 1971). Landscape as a scenic resource expanded from, *'... concern with the uniquely beautiful .... to the uniquely ugly... and finally to the everyday, non-unique landscapes in which most (people) live, recreate, work and travel'* (Zube et al. 1974, vii).

Landscape quality assessment presents a conundrum. Planners, geographers and others classify landscape on inherent physical qualities established with certain assumptions and criteria. Personal preferences do not occur, e.g. Linton (1968)

Fig. 3.4 Mountain scenery, Iceland

commented that mountains have high landscape quality (Fig. 3.4), resulting in landscapes being classified on a numerical, high, medium or low quality. The subjective basis of the selected criteria is ignored emphasising the role of scenery in attracting tourism. '*Landscape is assumed to be a quality present in the scene, a quality which one visits to see, experience and enjoy* (Lothian 1999, 178). The perception approach uses objective psychological testing and looks to community preferences in order to obtain an impression of the landscape, *i.e.* it eliminates the subjectivity of the researcher.

Many scenic assessments have been made based upon subjective analyses, but few acknowledge the implications of subjectivity on the validity of the results (Kaye and Alder 1999). For example, in an attempt to boost tourism, a quantitative method was used on the Central Coast region of Western Australia by assessing the presence of infrastructure, levels of attraction and environmental degradation (Priskin 2001), but no attempt was made to reduce the subjectivity of the observer, which reduced the results credibility. In the same vein Cocklin et al. (1990) assessed the potential of certain New Zealand areas for recreation possibilities, which graded sites on a High, Medium and Low scenic value basis. All were based on subjective observations of landform and cover plus presence of rare features.

After all:

What is the hardest thing of all?
That which seems the easiest:
For your eyes to see,
That which lies before your eyes. Goethe, quoted in Luijk (2012, 12)

Assessment of landscape scenery has evolved considerably during the past 30 years even though Appleton's (1975a) dictum still prevails, i.e. it exists in a theo-

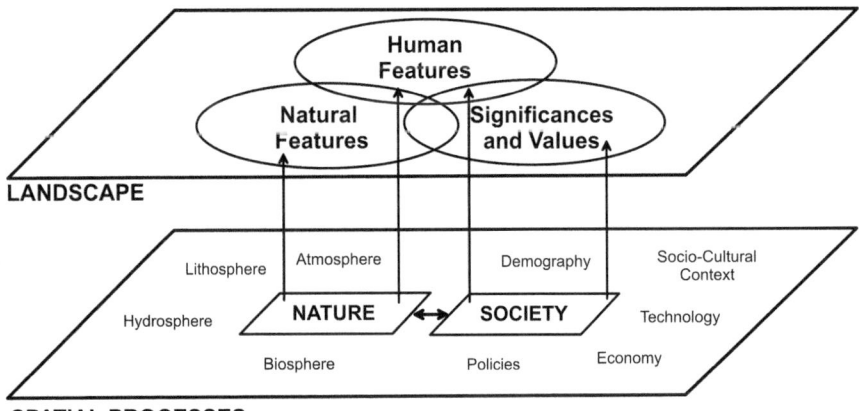

Fig. 3.5 A conceptual model for landscapes (after Castiglioni 1973, cited in Luijk 2012)

retical vacuum. Zube et al. (1982, 25) went further arguing that research without a general theory is fragmentary and has a hit-or-miss quality to it, and it was hard to understand how various research efforts fit together. Ideas that permeate the field have followed the pattern shown by Smith et al. (2002) for geomorphological research in landscape development in that they conform to the model for scientific change proposed by Kuhn (1962). Major quantum shifts characterise conceptual thinking, rather than a gradual diffusion of ideas and 'falsificationism' as advocated by Popper (2005). Brady (2003) has given an excellent assessment of these ideas. Many checklists and photographic inventories have been utilised, and important concepts expressed, for example by, amongst many others, Fines (1968), Leopold (1969), Shuttleworth (1980a, b), Penning–Rowsell (1989), Penning-Rowsell and Lowenthal (1986), CCW (1996, 2001), TLE (2013).

This twofold landscape approach entails a basal fundament onto which is super-imposed the visual landscape, consisting of a three-pronged overlapping Fenneman (1919) system diagram. In essence, these are the natural, anthropogenic and signifi-cance/values, (Fig. 3.5) that are assigned to the observed features, the latter is of importance only to society, determining how people view landscapes and has three components:

- The aesthetic sphere.
- The effective sphere *i.e.* landscape is part of one's own identity.
- The symbolic sphere *i.e.* elements within the landscape are of particular signifi-cance to people who perceive them.

Landscape evaluation is strongly rooted in the man-environment tradition and *'scenery has become a resource'* (Linton 1968, 219) and it is an *'economic resource and not a dispensable luxury'* (Clark et al. 1984, 34), *'because it is an accepted component of resource assessment programmes'* (Kaye and Alder 1999, 302–304). Therefore managers need to attempt an evaluation of scenic resources in an objec-tive and quantitative manner (Priskin 2001). Developing a methodology to fit with the constraints of quantification entails measuring the contributions of specific land-scape elements to overall preference (Buhyoff and Riesenmann 1979) remembering that scenery is a vital component for coastal tourism and drives the economy of many coastal countries.

A sub-section of coastal tourism is beaches and these are worth billions of tourist dollars (Clark 1977) and considered a major player in the tourism market (Houston 2013) and many tourism oriented countries, e.g. around the Mediterranean rim, have developed proactive growth policies along the coastal area (Benoit and Comeau 2005). For example, in Spain, France, Italy and Greece tourism receipts account for some 5% of the gross domestic product (UNWTO 2008), these countries account-ing for *'the most significant flow of tourists. A sun, sea and sand (3S) market'* (Dodds and Kelman 2008, 58). California beach visits exceed 567 million/year compared to 286 million to all National Parks USA (Dodds and Kelman 2008). After questioning >200 beach users at Gower, UK, Morgan and Williams (1995) found that scenery was the first choice of prioritised beach aspects; Unal and Williams (1999), questioned 120 beach users at Cesme Peninsula, Turkey, and con-cluded that scenery ranked second after clean bathing water. Many designated areas, such as National Parks etc., all reflect the scenic component in advertising the beauty of the area.

National and cultural differences in landscape appreciation and preference within the literature are sparse, especially with regard to coastal landscapes. Eletheriadis et al. (1990) found agreement between European nationality groups with regard to the most/least preferred landscapes – but also many significant differences, which were attributed to cultural differences and home environment familiarity. Lyons (1983) found that preferences changed throughout the life cycle: 6–12 year olds were more enthusiastic and less consistent than others were; 12–20 year olds gave a dip in preference levels generally, which rose slightly after the age of 20. From 36 years of age onwards, there was a slow, downward trend in preference levels. Preferences also changed with respect to gender, sex and residence.

Modern approaches have focussed on development of more detailed specific approaches For instance: Yamashita (2002) explored the scenic perception between adults and children. His conclusion was that it was the immediate surroundings that children focus upon, whereas adults were more aware of distal aspects, as well as the immediate surroundings; so do scenic evaluations adequately represent chil-dren's perceptions?

Fines (1968) and Kaplan and Kaplan (1989a, b) found natural landscapes to be perceived as more distinguished, spectacular and more scenic amongst culturally homogenous participants, suggesting that landscape preferences relate to landscape stimulation, as well as a need to make sense of it. Zube and Pitt (1981) and Zube and Sell (1986) showed that not all cultures shared the same perception of anthropo-genic landscapes, suggesting that man might be taught, or implicitly led to believe, that scenic beauty is an attribute of unmodified landscapes (Fig. 3.6). They sug-gested that environmental experience and landscape familiarity were very important factors in shaping perceptions, the opposite to that found in studies by, for example, Wellman and Buyhoff (1980), Buyhoff and Arndt (1981).

Fig. 3.6 Norwick is a tiny settlement in the Scottish archipelago of Shetland, located on the island of Unst. During World War II, the Royal Air Force built a Chain Home radarstation at Norwick. (Photo by Arthur Maxwell)

Zube et al. (1982, 25) pointed out that '*the first phase of landscape quality research was conducted in a more or less benevolent atmosphere (in both the U.S.A. and Britain) where scenic quality was assumed to be important by national leaders. That phase is ended. To continue in the same theoretical way in present research is to risk loss of consideration of scenic values in future landscape policy decisions.*' Have modern policy decisions followed this comment, as in order for scenery to take a prime position in political decision making, it must involve making a significant contribution to improving economic/social factors.

Areas of high ecological value usually coincide with areas considered to have significant natural beauty and Lee et al. (1999) proposed that it was possible to assess landscape quality based on vegetation cover evaluation alone. Canters (2002) and Gulinck et al. (2001) documented similar techniques predicting measurable land cover indicators that corresponded with areas of high scenic value. GIS imagery has been to the forefront of modern approaches and in New Zealand, Henderson and de Lambert (1992) suggested that utilisation of GIS overlays on land types be used in conjunction with other assessment techniques. New Zealand scenic evaluation utilises assessment criteria (cultural, historical, etc.) as surveys to identify coastal areas of significant natural character, which makes them less useful for determining strategies for the protection of scenery alone (Houlahan and Findlay 2004).

These examples – and see later ones – highlight the increasingly technical nature of scenic evaluation but fail to address simple issues, such as, variations in people's perception, or the basic inaccuracies in analysing landscapes based on images. Land cover mapping can give a general impression of a site but ground based fieldwork should be incorporated.

Yeomans (1986) suggested that when individual experts are the primary determinants of scenic evaluation the reliability issue could be ameliorated by repeating assessments made by others and cross checking results, as the aesthetic experience of individuals range considerably. Smith and Theberge (1987) highlighted the need to re-evaluate any technique based on assumptions, as perceptions vary over space through time The Landscape Institute (2011) argued that expert scenic assessment incorporates a balance of qualitative/quantitative judgment to increase credibility. Palmer and Hoffman (2001) researched the implications of a US Supreme Court decision requiring experts to provide reliable and valid assessments when providing evidence. They found that few reports contained such assessments and some that did had alarming results. Dakin (2003) commented that one should go beyond the use of experts and participation by the wider public and feedback of their perceptions helps to produce better environmental outcomes. In the USA, scenic assessment implementation has caused some concern and a shift has occurred from 'visual resource management' determined by expert based assessment to a more community focused and public participatory approach.

Many national bodies have incorporated scenic assessment studies as part of wider resource management strategies. The Australian strategy for locations with outstanding scenic quality was based on physical parameters and as a result, a number of coastal reserves have been established. This has been a difficult process due to the intersection of public and private land for management and planning (Harvey and Caton 2010).

In essence, scenic assessment has had a long and chequered history in that landscape descriptions appeared in many a Victorian explorer's notebook, as well as military manuals. A voluminous literature exists on this general topic, which rapidly shrinks when the coastal scene is involved, as formally coasts were primarily seen as having only ecological value (Sheail 1984). During World WII, a UK coastal scenery inventory and scientific interest assessment led in 1973/4 to the founding of the England and Wales Heritage Coast movement (Steers 1944). This originally had 27 designated areas but now has 45 and the organisation manages some 34% of the coastline of England and Wales, UK. with an emphasis on scenic beauty.

It was in the 1960–80s that a flurry of papers occupied themselves with the question of 'how does one assess scenery?' Many authors/authorities have written about the problems using a myriad of techniques, such as, Shafer et al. (1969), Fines (1968), Linton (1968, 1982), Leopold (1969), Shafer and Meitz (1970), Coventry-Solihull Warwickshire (1971), Robinson et al. (1976), Carlson (1977), Briggs and France (1980), Buyoff and Arndt (1981), Zube et al. (1982), Penning Rowsell (1982, 1989), Williams (1986), Countryside Commission (1987, 1993), CCW, (1996, 2001), CoE (2000), Ergin et al. (2002, 2006), TLE (2013). A summary for some of the above is given below.

3.3   Some Existing Scenic Methodologies: A Review

Though this be madness, yet there is method in't. W. Shakespeare, Hamlet, II, ii, 237

Evaluation techniques cover a spectrum from those based upon subjective landscape assessments by individuals/groups e.g. Shafer et al. (1969) or via a landscape's physical attributes e.g. Leopold (1969). Arthur et al. (1977) classified them into descriptive and public preference inventories, with quantitative and non-quantitative sub-sections for both models. The former has the largest category of techniques (quantitative and qualitative) for landscape assessment and these are usually applied by experts in an objective manner, e.g. Leopold (1969). A critique has been that landscape components are arbitrarily identified by experts, which is not the case in the checklist technique given in this book (Chaps. 4 and 5), as the 26 parameters identified were obtained as a result of many public questionnaire surveys. These surveys relate to Public Preference Models including psychological and phenomenological, e.g. Kaplan and Kaplan (1989a, b), Shuttleworth (1980a). Subjective assessment by many individual users can cover the range of perceptions involved for the total landscape and quality/beauty is discerned through judgement of these preferences (Jacques 1980). Other researchers have used different techniques to sub-divide assessments, e.g. direct and indirect (Briggs and France 1980), preference and surrogate components (Croft 1975), but it all revolves around experts vs. subjective public viewpoints, and coastal management must surely have an objective methodology in order to make meaningful decisions regarding any phenomena, and not only for scenery.

Fines (1968) linked photographs to landscape features in map grid squares and identified landscape units. The method involved evaluation of 20 'still' black and white photographs by a group of 45 representative people (what are these?) split into sub-groups according to sex, training and experience in a design discipline of each person. The mean value for each view was calculated for each sub-group and the highest mean value divided by the lowest gave a six category range of values, from unsightly (0–1) to spectacular (16–32). In the UK 18 was the highest obtainable value found in Highland areas; 12 in Lowland ones, with townscapes being ranked lower than natural landscapes.

Several disadvantages resulted from this study, mainly from lack of definitions and an assumption that all participant viewpoints were the same and more importantly it was assumed that photos are surrogates of scenery, a point not proved in the study. On this point it is interesting to note the following comment: '*An article with a definite tendency is for the most part read only by people who can already be reckoned to this tendency. At most a leaflet or a poster can, by its brevity, count on getting a moment's attention from someone who thinks differently. The picture in all its forms up to the film has greater possibilities. Here a man needs to use his brains even less; it suffices to look, or at most to read extremely brief texts, and thus many will more readily accept a pictorial presentation than read an article of any length. The picture brings them in a much briefer time, I might say at one stroke, the enlight-*

enment which they obtain from written matter only after arduous reading' (Hitler 1974, 427). It appears that Hitler was a fan of pictorial representation.

Shafer et al. (1969) and Shafer and Meintz (1970) used personal interviews and photographs (100, 8″ by 10″, black and white) in the USA to design an equation of landscape evaluation. Preferences scores were obtained *via* adding the resulting 50 rank values for each photo and the total for each photo was designated the preference score Y. From these they found that 10 significant terms explained 66% of the variation in landscape reference scores. They postulated six different landscape types based on: perimeters of immediate vegetation (individual leaves, not such fine detail, unable to distinguish vegetation shapes); and area (intermediate vegetation, water, distance of non-vegetation sector). The authors urged caution for resource planners using the technique.

Linton (1968) produced a composite assessment of scenery in Scotland and argued that appraisal was founded on scenic elements that influence one's reactions to it. He recognised two types.

Landform landscapes based upon relative relief, slope, abruptness, frequency, dissecting valley depth. He argued that processing landscape landforms was a tedious process, so substituted a subjective relief appraisal of features on 10″ to the mile OS maps, producing; mountains (8), bold hills (6), hills (5), plateaux (3), low uplands (2), lowlands (0).

Land usage – urban/industrialised (−5), continuous forest (−2), treeless farmland (1), moorland (3), varied forest/moors (4), richly varied farmed landscapes (5), wild (6).

Summation of the above bracketed numbers enabled a scenic resource map to be drawn. Relief was the dominant factor and the distance over which the view extended was not defined as was the differing quality of urban landscapes. This was very subjective and water was ignored.

Leopold's (1969, 4) main aim was to reduce subjectivity so that results, *'could be used in many planning and decision making context,'* carried out by ranking parameters (on a 1 to 5 scale *i.e.* bad to good) producing calculations/graphs, which rated chosen sites according to their aesthetics. This seminal paper by Leopold (1969) stressed the value of uniqueness of scenery and developed an index of river and landscape character based on physical, biological and human parameters using some western USA rivers, e.g. the Snake, as an example of the technique. The essence of his philosophy was that any landscape that was unique in either a positive/negative manner is of more significance than one that is common. He was well aware of the techniques shortcomings, e.g. a uniqueness score does not necessarily indicate attractiveness or unattractiveness, it does not differentiate between the significance of different factors, but was only one characteristic of the environment (what about intensity, novelty, etc.?). Some of Leopold's (1969) factors were directly measurable, others estimated and each observation was assigned to one of five attributes – see Chaps. 4 and 5. No weightings were given and each site parameter was assigned a uniqueness ratio. The validity of the scoring system cannot be proven but its use by previous observers has shown it to be consistent and operation-

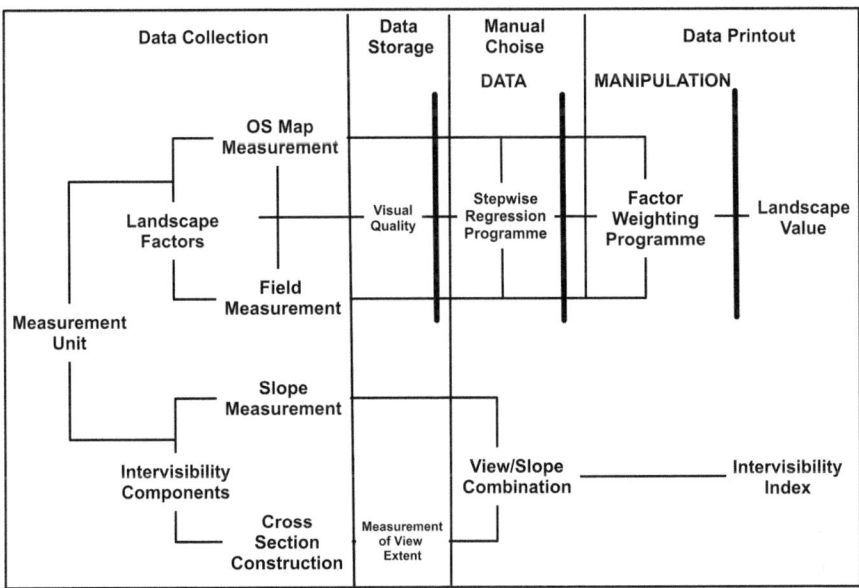

Fig. 3.7 Landscape evaluation technique. (Based on the Coventry Solihull Warwickshire Study, 1971–75, 123)

ally reliable. This was the seminal paper that resulted in the scenic assessment technique given in this book (Chap. 4). Although continued knowledge acquisition is necessary, there is an increasing need for innovative manipulation so his original checklist was amended.

The Coventry-Solihull-Warwickshire, UK (1971) study derived an aesthetic directory from assessment of landform, usage and land feature measurements culminating in an index of landscape value (Fig. 3.7). The base unit was the square kilometre and they delineated 24 factors, e.g. mining, water, railways, hedgerows, listed buildings, density, roads, farmland, residential, woodland. Specific examples being *landscapes worthy of conservation* (+60 to +80), e.g. parks (Arbury, Umberslade); *minimum acceptable residential areas* (−40), e.g. Shirley, Cranmore Industrial estates; *average urban areas* (−10), e.g. Otton, Tile Hill; *Landscapes comparable to Areas of Outstanding Natural Beauty* (+90 to +120), e.g. Weston Park.

Multiple regression analysis was performed to discover how far the variation between measured factors of 1 km^2 and those of another accounted for variation between visual quality scores subjectively given to those squares. This was one of the few techniques when weightings (derived from subjective surveys by the planners) were applied to each factor and these were multiplied by the incidence of the measured factors to give a landscape value, which ranged from −120 to +120 and represented the inherent quality of each square kilometre. Databases were set up in a GIS format and analysed via techniques, such as, TWINSPAN. Most sub-regions

fell in the farmland category but results closely corresponded with subjective surveys, not surprising given the high multiple correlation coefficients found. Home
interviews were also undertaken classified by Gross Rateable Value bands (£30 and
under, £31–56, £57–100 and £101 and over) and 117 responses were recorded with
12 refusals; houses empty/not located etc. amounted to 31.

Zube et al. (1974) working in the Connecticut River Valley, USA used 56 landscape settings of the area, which were deemed to give a representative coverage of
types, and identified 23 landscape dimensions as being significant determinants of
landscape quality. Each of these 56 settings was measured, using photos and topographic maps for: landform, area, land use area, edge and contrast, water and view,
and 307 people took part in the study. One hundred and twenty four correspondents
evaluated eight of the 56 sites and described them from a 51 item checklist together
with an 18, seven point (1–7), bi-polar semantic differential scale (varied), mountainous, common, unusual, tidy, closed, open, etc. Panoramic photographs of these
sites were then assessed as to scenic quality. Colour photos of all 56 sites were also
evaluated by a Q-sort procedure by the remaining 183. A strong correlation was
found between on-site analysis and photographs. There was an unexplained vari-
ance between physical elements and perceptual responses. A series of papers by
Buhyoff and Riesenmann (1979), Buhyoff and Wellman (1980) and Buhyoff and
Arndt (1981) later investigated the problems associated with dimensionality as significant determinants, culminating in Kroh and Gimblett (1992) who found little
relationship between live scenery viewing and pictures, arguing that multi-sensory
experiences were very important.

However, as regards photos being scenery surrogates, Williams and Lavalle
(1990) showed conclusively that this was perfectly acceptable if they viewed the
same scene taken with a 50 mm camera lens (NB the human eye field of vision is
145°). To date this has not been done. They took three groups of people (skilled,
semi-skilled and laymen; n = 45) to 28 coastal sites. Two pictures were shown: one
of 'heaven' (positive score + 100; serene, lovely coastal view) and one of 'hell'
(negative, score − 100; industrial coastal cement factory). A box was then fitted to
the observer's head so that the view was restricted to that of a 50 mm camera lens.
Observers were then asked to look at a scene, evaluate it and give a value with
respect to the two control pictures. Scores were then standardised and a colour
photo taken with a 50 mm lens of the exact scene. Two weeks later the same observers were shown photographs taken on the field trip and asked to repeat the scoring.
Statistical testing showed no differences between groups. Dupont et al., (2015) have
given an excellent review on this topic.

Penning-Rowsell (1989) assessed public attitudes and perception in the UK as
did Pendleton et al. (2001) in the USA. Penning-Rowsell (1989) investigated the
evaluation techniques of various UK County authorities, which involved five questions sent to 165 authorities of which 93 authorities had not used any assessment in
the past 5 years. Sixty-seven percent replied and 22% of these used some type of
systematic evaluation procedure, while 13% indicated reservations of the techniques used.

With the rise of GIS, Bishop and Hulse (1994) using a raster data base of a 6 km square in Oregon, and Brabyn (1996), demonstrated the potential of GIS for analysis of scenic beauty. IEMA (2011) showed that the iterative design approach to Environmental Impact Assessment (EIA) was commonly recognised and its value widely accepted by practitioners.

In essence, scenic evaluation emerged mainly in the mid twentieth century, as a tool to separate the classification and description of landscapes. It was the process of identifying the importance of a particular landscape, landscape type or landscape feature, by reference to specified value criteria, i.e. it is one form of landscape appraisal to identify value criteria against which value can be determined.

In the twenty-first century this morphed into Landscape Character Assessment (Fig. 3.8), defined as, '*a distinct, recognisable and consistent pattern of elements in the landscape that makes one landscape different from another, rather than better or worse*' (LI. 2016, 7) of which scenery plays but a part. It identifies/explains the unique combination of elements and features (characteristics) that make landscapes distinctive – and this is still evolving (Daniel 2001; Swanwick 2002; Tudor 2014; Saeidi et al. 2017).

Earlier UK Countryside Commission (CC) reports (1987, 1993) confirmed that landscape character assessment factors were natural (e.g. geology, soils), cultural (e.g. land use, settlements), aesthetic (visual, sounds) and associations (e.g. history). In Wales, UK these have all been incorporated into the LANDMAP (Landscape Assessment And Decision Making Process) system, via, a partnership programme between the Countryside Council for Wales (now Natural Resources Wales) and all the Unitary and National Park Authorities of Wales (CCW 2001). It was a benchmark publication and a tool for assessing landscape resources to aid decision-making and natural resource planning at a range of levels (local to national) where landscape characteristics, qualities and influences on the landscape are recorded. The scales utilised are commonly:

- National and Regional, typically at 1:250,000 (broad patterns in landscape character variation).
- Local Authority, usually at county, unitary authority or district level, 1:50,000 or 1:25,000 (identifying landscape types and/or areas.)
- Local or Site Level, normally at 1:10,000 or larger scales.

The core of LANDMAP comprised five spatially related datasets recording information about the characteristics, qualities that influence the geology, habitats, visual and sensory, historical and cultural on the landscape, which is managed through a complete Geographical Information System. The areas are represented as single polygons (or polyline) in a single map layer. Therefore, LANDMAP is represented by five map layers in a GIS format. Also utilised were aerial photographs, preferably orthographically corrected, and available as continuous digital coverage for a study area (not as individual tiles) and local/unitary development plans/documents, e.g. landscape assessments. A hierarchical classification system was utilised aiming to classify landscape into areas of distinct Visual & Sensory character at four separate levels (Table 3.1). The landscape qualities/characteristics of discrete areas are first mapped and survey records and management recommen-

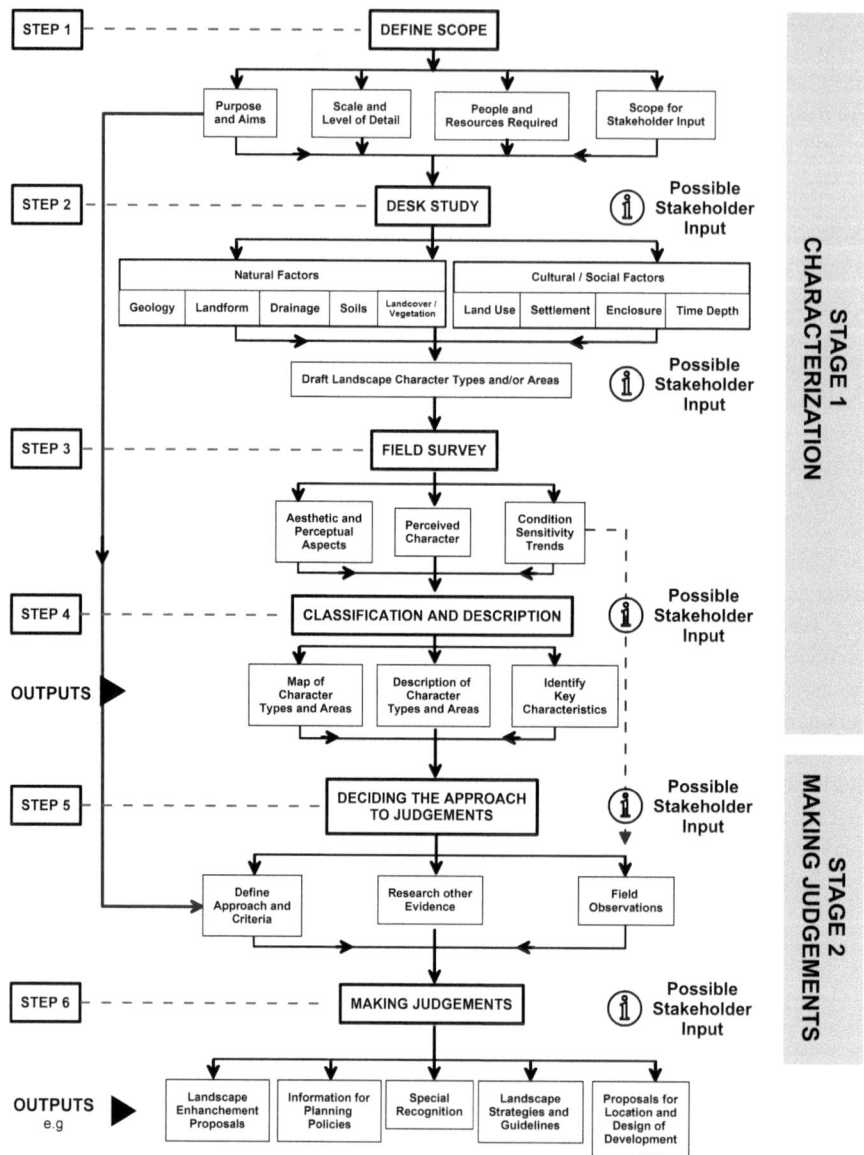

Fig. 3.8 Landscape character assessment. (After Swanwick 2002, 13)

dations and criteria based evaluations added. The terminology differs slightly from that of Landform Character Assessment (LCA) and Fig. 3.9 indicates the relationship.

Overlays can now be produced utilising remote-sensed satellite data, showing each variable on an ordinal scale, which can be combined to identify various feature sets (physical characteristics and land cover – Fig. 3.10).

Table 3.1 Hierarchy levels for landmap

Level 1	Level 2	Level 3	Level 4
Broad landform and land cover	Landform	Land cover	Detail – Location/scale/exposure/settlement

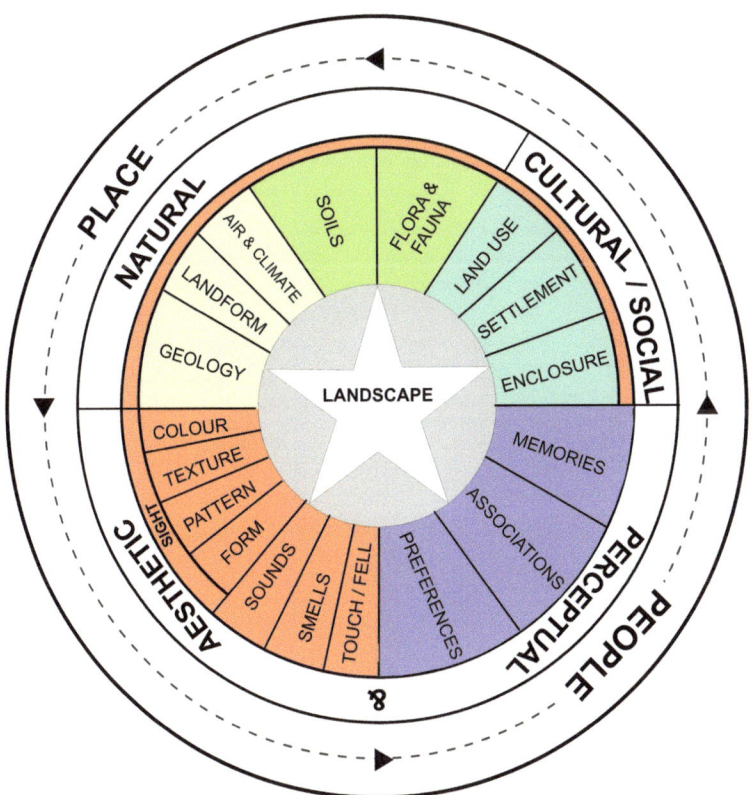

Fig. 3.9 The relationship between Landform and LCA. (After Tudor 2014)

Landscape Character Assessment is now being increasingly used to inform urban, or townscape, assessments, together with Seascape Character Assessments (Briggs et al. 2011; NE 2012; Fig. 3.10). The Marine and Coastal Access Act 2009 (National Archives 2009) is a policy statement which specifically established the concept of 'Seascape' i.e. landscapes with coastal/seas views, and coasts together with the adjacent marine environment which have, natural, cultural, historical and archaeological links with each other (NE 2012).

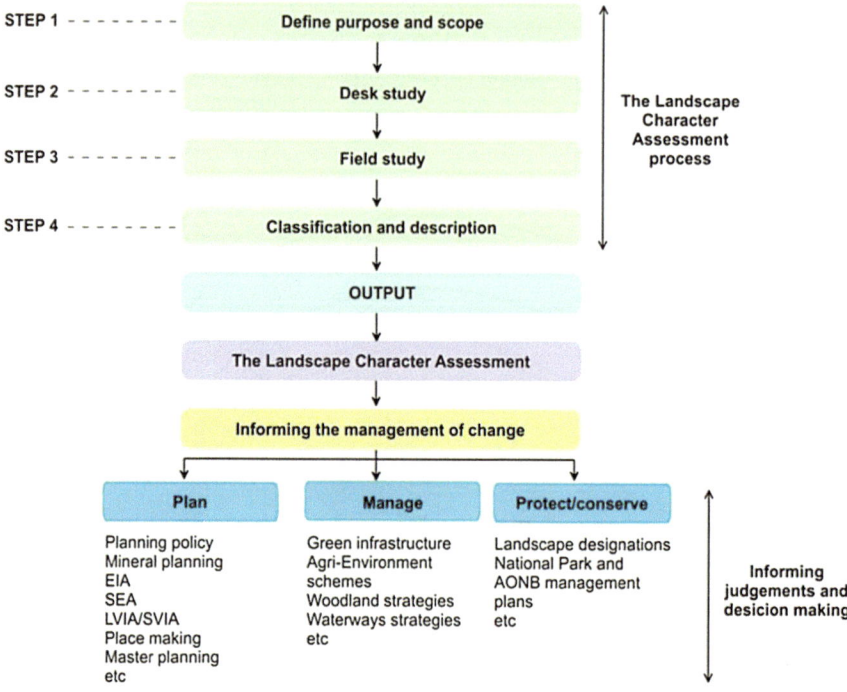

Fig. 3.10 Landscape Character Assessment and making judgements. (From NE 2014)

3.4 Conclusions

> *A wise man that had it for a byword, when he saw, men hasten to a conclusion.* Francis
> Bacon, Of Dispatch, 25

Numerous scenic assessment techniques are in existence and a few have been enumerated in this chapter. Virtually all checklists used to quantify scenery or beaches have two major failings: no weighting of parameters, e.g. the annual Dr. Beach ratings of USA beaches (Leatherman 2016) and no viewers preferences are taken into account, it is usually the 'experts' opinion. All form a spectrum where extreme techniques are based on subjective assessments of landscape quality by individuals or groups and at the other extreme, by methods, which use physical attributes of landscape as surrogates for personal perception. The essence of attention is that it is elective. *'We can focus on something in our field of vision, but never on everything. All attention must take solace against a background of inattention…. To see all, we must isolate and select'* (Gombrich 1965, 515). So it seems as if scenic beauty can be viewed rather as one would regard a Rorschach ink blot test!

Dearden (1985) argued that there is seemingly a considerable diversity of philosophy and approach in published landscape assessment papers/books, even though all were concerned with understanding landscape and its visual quality, concluding that landscape quality is not a simple concept and cannot be fully appreciated through application of one universal approach or technique. It is NOT simple but some 30 years later much progress has been made. The technique given in Chap. 4 (scenic parameter selection, weightings, fuzzy logic to deal with uncertainties), is an attempt to synthesise the various strands of research mentioned above with respect to scenic analysis.

However, the basic theme of Appleton's (1975a) seminal paper still holds true today, i.e. there are many practitioners and techniques for landscape assessment but still a theoretical vacuum exists. Based on the above, Appleton's (1975b) classic book represented a search for landscapes that would incite an aesthetic response to man and covered a spectrum ranging from philosophy to ethology. Dewey (1929, 14) argued that *'a theory of the place of the aesthetic does not have to lose itself in minute details when it starts with experience in its elemental from. Broad outlines suffice.'* We still await the connecting syntax between experience and landscape, remembering that *'connecting with immediate surroundings, through tactile and sensory engagement, is so basic and constant'* (Clark and Clark 2009, 311); or as Berger (1972) put as his final comment *'all should be gauged against your own experience.'*

Therefore the relationship between what one sees and what one knows is seemingly not settled. However, the final words of this chapter should be those of one of the great pioneers in the landscape field, one who posed the question, 'what' does one like in landscape and 'why?' *'It is true that the theory underlying the judging of a fine wine or a good piece of sculpture is probably as obscure as that which underlies the evaluation of landscape. In those arts, however, we still have some faith – possibly misplaced – in the ability of the expert to recognise excellence, however defined'* (Appleton 1975a, 122).

References

Appleton J (1975a) Landscape evaluation: the theoretical vacuum. Trans Instit Brit Geog 66:120–123

Appleton J (1975b) The experience of landscape. Wiley, London

Arthur LM, Daniel TC, Boster RS (1977) Scenic assessment: an overview. Landsc Urban Plan 4:109–129

Benoit G, Comeau A (2005) Méditerranée. Les perspectives du Plan Bleu sur l'environnement et le développement. Editorial de l'Aube, Paris

Berger J (1972) https://www.youtube.com/watch?v=5jTUebm73IY

Bishop ID, Hulse DW (1994) Prediction of scenic beauty using mapped data and geographic information systems. Landsc Urban Plan 30:59–70

Brabyn L (1996) Landscape classification using GIS and National Digital Databases. Landsc Res 27:277–300

Brady E (2003) Aesthetics of the natural environment. Edinburgh University Press, Edinburgh

Briggs DJ, France J (1980) Landscape evaluation: a comparative study. J Environ Manag 10:263–275

Briggs J, Campbell K, Houlston I, Tudor C (2011) Seascape character assessment guidance for England, Scotland & Wales: Consultation Draft V1.2A. Natural England Commissioned, London

Buhyoff GJ, Arndt LK (1981) Interval scaling of landscape preference by direct and indirect measurement methods. Landsc Plan 8:257–267

Buhyoff GJ, Riesenmann MF (1979) Experimental manipulation of dimensionality in landscape preference judgements: a quantitative validation. Leis Sci 2:221–238

Buhyoff GJ, Wellman JD (1980) The specification of a non-linear psychophysical function for visual landscape dimensions. J Leis Res 12:257–252

Canter F (2002) Assessing the effects of input uncertainty in structural landscape. Inst J Geog Inform Sc 16(2):129–149

Carlson AA (1977) On the possibility of quantifying scenic beauty. Landsc Plan 4:131–172

CC – Countryside Commission (1987) Landscape assessment- a countryside commission approach. CCD, Cheltenham

CC – Countryside Commission (1993) Countryside commission, landscape assessment guidance. CCD, Cheltenham

CCW – Countryside Council for Wales (1996) The Welsh landscape: our inheritance and its future protection and enhancement. Countryside Council for Wales, Bangor

CCW (2001) The LANDMAP information system, 1st edn. Countryside Council for Wales, Bangor

Clark JR (1977) Coastal ecosystems management: a technical manual for the conservation of coastal zone resources. Wiley-Interscience, New York

Clark E, Clark TL (2009) Isolating connections – connecting isolations. Geogr Ann Ser B 91:311–323

Clark BD, Gilad A, Bisset R, Tomlinson P (1984) Perspectives in environmental impact assessment. Springer, Amsterdam

Cocklin C, Harte M, Hay J (1990) Resource assessment for recreation and tourism: a New Zealand example. Landsc Urban Plan 19:291–303

CoE (2000) Council of Europe European landscape convention. Council of Europe, Florence

Countryside Act of 1968, Chapter 41: printed in England by Harry Pitchforth, controller of her Majesty's Stationery Office and Queen's Printer of Acts of Parliament, 64pp (371998)

Coventry-Solihull-Warwickshire (1971) A strategy for the sub-region. Supplementary report 4 – evaluation. GVA Grimley, London

Croft RS (1975) The landscape component approach to landscape evaluation. Trans Inst Brit Geogr 66:124–112

Dakin S (2003) There's more to landscape than meets the eye: towards inclusive landscape assessment in resource and environmental management. Can Geogr 47(2):185–201

Daniel TC (2001) Visual landscape planning in the 21st century. Landsc Urban Plan 54(1–4):267–281

Davidson J, Wibberley G (1977) Planning and the rural environment. Pergamon Press, Oxford

Dearden P (1985) Focus on landscape aesthetics. Can Geogr 29(3):262–273

Dewey J (1929) Experience with nature. W W Norton & Co, New York

Dodds R, Kelman I (2008) How climate change is considered in sustainable tourism policies: a case of the Mediterranean Islands of Malta and Mallorca. Tour Rev Int 12(1):57–70

Dupont L, Antrop M, Van Eetvelde V (2015) Does landscape related expertise influence the visual perception of landscape photographs? Implications for participatory landscape planning and management. Landsc Urban Plan 141:68–77

Eckbo G (1969) The landscape we see. McGraw Hill Inc, New York

Eletheriadis N, Tsalikidis I, Manos B (1990) Coastal landscape preference evaluation. A comparison among tourists in Greece. Environ Manag 14(4):475–487

Ergin A, Williams AT, Micallef A, Karakaya ST (2002) An innovative approach to coastal scenic evaluation. In: Ozhan E (ed) Beach management in the Mediterranean and Black Sea. Medcoast/METU, Ankara, pp 215–226

Ergin A, Williams AT, Micallef A (2006) Coastal scenery: appreciation and evaluation. J Coast Res 22(2):958–964

Faulks S (1999) Charlotte Grey. Hutchinson, London

Fenneman NM (1919) The circumference of geography. Ann Assoc Am Geogr 9(1):3–11

Fines KD (1968) Landscape evaluation. A research project in East Sussex. Reg Stud 2:41–55

Gillespie W (1970) Landscape evaluation. Seminar paper. National countryside classification structure. Landscape Research Group, Oxford

Gombrich E (1965) Visual discovery through art. In: Wilkinson J, Hogg R (eds) Theories of art and beauty. Open University Press, Lincoln, pp 505–521

Greenbie BB (1975) Problems of scale and context in assessing a generalized landscape for particular persons. In: Zube EH, Brush RO, Fabos JG (eds) Landscape assessment: values, perceptions and resources. Wiley, New York, pp 65–91

Gulinck H, Mugica M, de Lucio JV, Atauri JA (2001) A framework for comparative landscape analysis and evaluation based on land cover data, with an application in the Madrid region (Spain). Landsc Urban Plan 55(4):257–270

Hartshorne R (1939) The nature of geography: a critical survey of current thought in the light of the past. Assoc Am Geogr 29(3):173–412

Harvey N, Caton B (2010) Coastal management in Australia, OUP, 342pp

Henderson E, de Lambert R (1992) Landscape values and resource management, report to Min. for the Environment, Wellington, New Zealand

Hitler A (1974) Mein Kampf, Hutchinson, 629pp

Houston JR (2013) The economic value of beaches – a 2013 update. Shore Beach 81(1):3–11

Houlahan J, Findlay S (2004) Estimating the "critical" distance at which adjacent landuse degrades wetland water and sediment quality. Landsc Ecol 19(6):677–690

Hungerford MW (1889) Molly Bawn. Dodo Press, London

IEMA (2011) Special report – the state of environmental impact assessment practice in the UK. IEMA, London

Jacques DL (1980) Landscape appraisal: the case for a subjective theory. J Environ Manag 10:107–113

Kaplan R, Kaplan S (1989a) The experience of nature: a psychological perspective. Cambridge University Press, Cambridge

Kaplan R, Kaplan S (1989b) The visual environment: public participation in design and planning. J Soc Issue 45(1):59–86

Kaye R, Alder J (1999) Coastal planning and management. E & FN Spon, London. 375pp

Kroh PD, Gimblett RH (1992) Comparing live experience with pictures in articulating landscape preference. Landsc Res 17(2):5869

Kuhn TS (1962) The structure of scientific revolutions. University of Chicago Press, Chicago. 217pp

Leatherman SP (2016) www.drbeach.org

Lee JT, Elton MJ, Thompson S (1999) The role of GIS in landscape assessment: using inter-based criteria for an area of the Chiltern Hills area of outstanding natural beauty. Land Use Policy 16:23–32

Leopold LB (1969) Quantitative comparisons of some aesthetic factors among rivers. US. Geological Survey, Circ. 620. Washington, DC

Lewis PF, Lowenthal D, Yi-Fu T (1973) Visual blight in America. Association of American Geographers, Washington, DC

LI – Landscape Institute (2016) Landscape character assessment technical information note 08/2015 February, 18pp

Linton DL (1968) The assessment of scenery as a natural resource. Scott Geogr Mag 84:219–238

Linton DL (1982) Visual assessments of natural landscapes. W Geogr Ser 20:97–116

Lothian A (1999) Landscape and the philosophy of aesthetics: is landscape quality inherent in the landscape or in the eye of the beholder? Landsc Urban Plan 44:177–198

Luijk A (2012) Aesthetics as a way of knowing... or experiencing. Oslo, 9th January

Lyons E (1983) Demographic correlates of landscape preference. Environ Behav 15(4):487–511

Macaulay (2014) Review of existing methods of landscape assessment. www.macaulay.ac.uk/ccw/task-two/evaluate.html. Accessed 30 Aug 2016

Marine and Coastal Access Act 2009 (2009) Legislation.gov.uk. www.legislation.gov.uk/ukpga/2009/23/contents

Morgan R, Williams AT (1995) Socio-demographic parameters and user priorities at Gower Beaches, UK. In: Healy MG, Doody JP (eds) Directions in European Coastal Management. EUCC & Samara Publishing, London, pp 83–90

National Archives (2009) http://www.legislation.gov.uk/ukpga/2009/23/contents

NE (2012) Natural England. An approach to seascape character assessment. NECR

NE (2014) Natural England, An approach to landscape character asessment. NECR

Olwig K (2003) Landscape: the Lowenthal legacy. Ann Assoc Am Geogr 93(4):871–877

Palmer JF, Hoffman RE (2001) Rating reliability and representation validity in scenic landscape assessments. Landsc Urban Plan 54:149–161

Pendleton L, Martin N, Webster DG (2001) Perceptions of environmental quality: a survey study of beach use and perceptions in Los Angeles County. Mar Pollut Bull 42(11):1155–1160

Penning-Rowsel EC (1982) A public preference evaluation of landscape quality. Reg Stud 16:97–112

Penning-Rowsell EC (1989) Landscape evaluation in practise – a survey of local authorities. Landsc Res 14(2):35–37

Penning-Rowsell EC, Lowenthal D (1986) Landscape meanings and values. HarperCollins Publishers, London, 137pp

Popper K (2005) The logic of scientific discovery. Routledge, London, 545pp

Priskin J (2001) Assessment of natural resources for nature-based tourism: the case of the Central Coast Region of Western Australia. Tour Manag 22(6):637–648

Relf EC (1976) Places and placelessness. Pion, London

Robinson DG, Laurie IC, Wager JF, Traill AL (1976) Landscape evolution: the landscape evaluation research project 1970–75. Manchester University, Manchester

Saeidi S, Mohammadzadeh M, Salmanmahiny A, Mirkarimi SH (2017) Performance evaluation of multiple methods for landscape aesthetic suitability mapping: a comparative study between multi-criteria evaluation, logistic regression and multi-layer perceptron neural network. Land Use Policy 67:1–12

Sewell WRD, Foster HD (1971) Environmental revival- promise and performance. Environ Behav 3:123–134

Shafer EL, Hamilton JF, Schmidt EA (1969) Natural landscape preferences; a predictive model. J Leis Res 1(1):1–19

Shafer EL, Meitz L (1970) It seems possible to quanify scenic beauty in photographs. United States Department of Agriculture Forest Service Research Paper NE-162

Shafer, Elwood L., Jr., and Mietz, James. 1970. It seems possible to quantify scenic beauty in photographs. *USDA Forest Service Research Paper NE-162. Upper Darby, Pa.* Northeastern Forest Experiment Station

Sheail J (1984) Nature Reserves, National Parks and post war reconstruction in Britain. Environ Conserv 11:29–34

Shuttleworth S (1980a) The evaluation of landscape quality. Landsc Res 5:14–20

Shuttleworth S (1980b) The use of photographs as an environmental presentation medium in landscape studies. J Environ Manag 11:61–76

Smith PGR, Theberge JB (1987) Evaluating natural areas using multiple criteria: theory and practice. Environ Manag 11(4):447–460

Smith BJ, Warke PA, Whalley WB (2002) Landscape development, collective amnesia and the need for integration in geomorphological research. Area 33(4):409–418

Starr G (2013) Feeling beauty: the neuroscience of aesthetic experiences. MIT Press, London

Steers JA (1944) Coastal preservation and planning. Geogr J 104:7–27

Swanwick C (2002) Landscape character assessment guidance for England and Scotland. The Countryside Agency and Scottish Natural Heritage, London

Teale EW (1966) Wandering through winter. Dodd, Mead and Co, New York

The Landscape Institute (2011) Guidelines for landscape and visual assessment, 3rd edn. 145pp

TLE (2013) The landscape institute. Guidelines for landscape and visual impact assessment. Landscape Institute, London

Tuan YF (1977) Space and place: the perspectives of experience. University of Minnesota Press, Minneapolis. 235pp

Tudor C (2014) An approach to landscape character and assessment. Natural England, London

Turner J (1987) Application of landscape values: a planner's view. Trans Inst Br Geogr 66:156–161

Unal O, Williams AT (1999) Beach visits and willingness to pay: Cesme peninsula, Turkey. In: Ozhan E (ed) Land Ocean interactions: monitoring coastal ecosystems. MEDCOAST – Middle East Technical University, Ankara, pp 1149–1162

UNWTO (2008) United Nations world tourism organization: tourism highlights. UNWTO, Madrid

Wellman JD, Buyhoff GJ (1980) Effects of regional similarity of landscape preferences. J Env Manag 11:105–110

Williams AT (1986) Landscape aesthetics of the river Wye. Landsc Res 11(2):25–30

Williams AT, Lavalle CD (1990) Coastal landscape evaluation and photography. J Coast Res 6(1):1011–1020

Yamashita S (2002) Perception and evaluation of water in landscape: use of photo projection method to compare child and adult residents' perceptions of a Japanese of a water environment. Landsc Plan 62:3–17

Yeomans WC (1986) In: Smardon RC, Palmer JF, Felleman JP (eds) Foundations for visual project analysis. Wiley, New York

Zube EH, Pitt DG (1981) Cross cultural perceptions of scenic and heritage landscapes. Landsc Plan 8:69–87

Zube EH, Sell JL (1986) Human dimensions of environmental change. J Plan Lit 1(2):162–117

Zube EH, Pitt DG, Anderson TW (1974) Perception and measurement of scenic resources in the Southern Connecticut river valley. Institute for Management and Historic Environment, Amherst. No. R-74-1, 171pp

Zube EH, Sell JL, Taylor JG (1982) Landscape perception: research, application and theory. Landsc Plan 9:1–33

Chapter 4
Coastal Scenery Assessment by Means of a Fuzzy Logic Approach

Ayşen Ergin

Abstract Landscape is a major element affecting people's life quality and coastal landscape evaluation is strongly rooted in the man-environment tradition. Coastal areas, all over the world, are under threat due to the conflicting requirements that rely on natural scenery of such as habitation, recreation, and industry. Since 'coastal scenery' is a natural resource, it has to be evaluated in an objective and quantitative way to provide a means of comparison against coastal activities and for environmental impact assessments. This chapter presents an evidence-based methodology called **'Coastal Scenic Evaluation System (CSES)'**. It is a technique that can be used not only for landscape preservation and protection, but also as scientific tool for envisaged coastal management and future development based upon plans formulated by an evidence-based approach. The results provide base-line information for a sound coastal management decision especially regarding intensive urban and industrial developments. CSES uses fuzzy logic to reduce subjectivity on decisions and obtain a quantitative evaluation of public survey research on 26 coastal scenic parameters having both physical and human perceptual characteristics. The weights of the scenic parameters were estimated by public survey questionnaires for Turkey, UK, Malta and Croatia and *via* consultations with coastal experts from the above mentioned four countries and Australia, Ireland, USA and Japan. Fuzzy logic mathematics was used to calculate a coastal scenic evaluation index (D) from the checklist of 26 scenic parameters by using the attributed weights of the parameters which enabled to categorize scenic values of the coastal areas into five distinct classes.

4.1 Quantifying Beauty

If one way be better than another, that you may be sure is Nature's way. (Aristotle)

A. Ergin (✉)
Department of Civil Engineering, Coastal Engineering Division, Middle East Technical University, Ankara, Turkey
e-mail: ergin@metu.edu.tr

© Springer International Publishing AG, part of Springer Nature 2019   67
N. Rangel-Buitrago (ed.), *Coastal Scenery*, Coastal Research Library 26,
https://doi.org/10.1007/978-3-319-78878-4_4

The world is facing critical environmental threats, therefore protection of the environment and conservation of natural resources in both local and global scale is becoming more and more prominent. Starting with the late 1960s and 1970s the range of environmental concerns broadened to include natural scenic beauty, wilderness experience, and outdoor recreation opportunity and visibility values and scenery defined as *"a picturesque view or landscape"* (Merriam-Webster Dictionary Online 2004) became a core scientific material. Evaluating scenery as a resource in an objective and quantitative fashion then became a challenging attempt for the scientist. From an ecological viewpoint, it was also stated that *"landscape is seen as a reflection of the mutual interactions between living organisms and their environment, considering landscape as an ecosystem"*, Van der Meulen (personal communication, 1997).

Many authors have written about scenic assessments stating that landscape evaluation classifications, in essence, includes expert-based approaches that rely on evaluations by professionals; perceptual and experimental approaches that obtain observer responses to scenery where identification often focuses on physical landscape features and human -environmental interaction in addition to aesthetics (for more information see Chaps. 2 and 3).

Amongst the many models and rating schemes that have accrued in this field, most have been in existence for *circa* 50 years. The more important evaluations, in chronological order, have been the works of Lowenthal (1967, 1978), Fines (1968), Linton (1968, 1982), Leopold (1969), Appleton (1975), Robinson (1976), Carlson (1977), Briggs and France (1980), Buyoff and Arndt (1981), Penning-Rowsell (1982, 1989), Kaplan and Kaplan (1989), Daniel (1990), Elettheriadis et al. (1990), Dakin (2003) and the authorities: the Countryside Commission (CC 1987, 1993), the Countryside of Wales (CCW 1996, 2001) as presented in Chap. 3.

Important techniques on scenic assessment, − those pertaining to field based objective replication studies on scenic assessment – can be briefly listed as (Linton 1968), statistical techniques obtained from site observations (Clamp 1976) and assessing public attitudes and landscape preferences (Penning-Rowsell 1989). Today, more scientific and professional disciplines are interested in environmental issues for management and protection studies than ever.

Recent studies have evaluated the physical features and utilization of coastal areas by taking into consideration human-nature interaction. The Countryside Commission (CC 1993) obtained a range of landscape types from assessing the natural landscape, cultural, aesthetic and associations. A similar approach, the LANDMAP series was devised for the Welsh landscape as LANDMAP Information System of Countryside Council for Wales (CCW 1996, 2001), which uses a geographical information system (GIS) which records, and makes available information about landscape qualities into a four level data set. The data set consists of contextual layers (landscape form, landscape function), GIS layers of core information (earth science, biodiversity, visual & sensory, history & archaeology, cultural), public perception. In any coastal scenic evaluation methodology, efforts to understand the effects of public perception are critical since public surveys used in the methodology depend on the environmental information received and processed by respondents, such as, the national and cultural background, age, gender, education, and training of the respondents which are highly subjective.

In scenic evaluation with regard to national and cultural differences there are few public perception studies. In view of this objective a British Council Project was carried out as the first pilot project of its type for Mediterranean trans-national boundaries with the participation of three countries (Turkey, UK and Malta) between 2001–2004. With respect to selected coastal sites in Turkey, UK, and Malta the main objectives of the project were:

- To assess coastal scenic quality by selected components based on scientific methodologies.
- To provide baseline information so that a sound scientific basis can be available for any envisaged subsequent management plans.

An innovative methodology '**Coastal Scenic Evaluation System (CSES)**' was developed based on expert and trained groups via public survey evaluation processes and published in British Council Reports (BCR 2003; 2004), Ergin et al. (2002, 2004, 2006, 2011). Within the methodology, coastal scenic evaluation comprised an assessment (priorities and preferences) of physical and human parameters which were addressed adequately by adapting fuzzy-logic mathematics to provide an evidence–based approach for sound coastal management decisions. During the project program, field work was carried out at selected 57 sites in Turkey, UK and Malta. Reports of these visits and the related studies are presented in BCR (2003, 2004). A pilot study for coastal scenic evaluation of some Dalmatian coasts was also undertaken in Croatia during the British Council Project studies and 33 coastal areas were assessed (Ergin et al. 2004). Coastal scenic assessment results of 90 coastal areas from the above stated four countries are presented in this chapter. Over a time span of a decade or so, 952 global locations were assessed by the authors of this book using the CSES. Coastal scenic assessment results of these are given in Chap. 5.

CSES evaluates the coastal scenery quantitatively, using fuzzy logic analysis to score the weighted coastal scenic parameters (18 physical and 8 human) by expert and trained groups via public surveys. CSES provides not only an objective evaluation of the physical parameters but also human parameters (adverse impacts of human–nature interaction; such as unattractive urbanization; intensive and not environmental friendly developments). The CSES also aims to point out how scenic parameters may be improved by judicious intervention relating to parameters which are mainly human.

4.2   Methodological Approach

So far as the laws of mathematics refer to reality, they are not certain.
And so far as they are certain, they do not refer to reality. (Albert Einstein, Geometry and Experience, 1921)

A five scale attribute rating system for the evaluation of 18 physical and 8 human coastal scenic parameters are defined for the "Coastal Scenic Evaluation System" (Table 4.1). Relevant definitions and pictures of the coastal scenic parameters are given in Chap. 5.

The main steps of the methodology are:

- The importance of each parameter and its parameter weights were determined by public perception surveys.
- A mathematical model based on fuzzy logic mathematics (Zadeh 1965) was developed to integrate the parameters' weights into an expert-rating system of scenic sites.
- An evaluation index (D) was computed as a decision parameter for each coastal site from a fuzzy mathematical model.
- The sites evaluated are further classified into five classes according to the D values achieved.

4.2.1   Public Surveys for Coastal Scenic Parameters

Scenic parameters, which constitute the essence of coastal scenery, were selected by literature surveys and carrying out questionnaire surveys, interviews with coastal users and consultancy with coastal experts, professionals and academicians in Turkey, UK, Malta and Croatia (BCR 2003).

To compute the weights of the 26 scenic parameters a cross-cultural evaluation via public surveys were carried out in Turkey, UK, Malta and Croatia (BCR 2003) and in some master's thesis studies (Gezer 2004; Karakaya 2004; Uçar 2004). Among 543 inquiries 58 were excluded (unmarked and/or double marked parameters) leaving 485 inquiries as given in Table 4.2 (Ergin et al. 2011) for the statistical analysis to compute the weights.

In Turkey, UK, Malta and Croatia, in public surveys for each scenic parameter every respondent was asked to give a rating on a five-point scale (from 1 to 5) representing the least important to the most important, respectively. They were also asked to list their top five most important parameters, which was used as an easy tool for viewing the most important parameters as presented with stars in the last column of Table 4.3.

Scenery appreciation is highly subjective and depends on a number of factors such as national and cultural background, age, gender, education. As described by Lowenthal (1967, 153) *'the environment is responded to and affected indirectly through the medium of a personally apprehended milieu. This milieu differs for each*

Table 4.1 Coastal Scenic Evaluation System (Ergin et al. 2004)

N	Physical Parameters		1	2	3	4	5
1	Cliff	Height (H)	Absent (< 5 m)	5 m ≤ H < 30 m	30 m ≤ H < 60 m	60 m ≤ H < 90 m	H ≥ 90 m
2		Slope	< 45°	45° – 60°	60° – 75°	75° – 85°	Circa vertical
3		Special features[a]	Absent	1 special feature	2 special features	3 special features	Many > 3 special features.
4	Beach face	Type	Absent	Mud	Cobble/boulder	Pebble/gravel	Sand
5		Width (W)	Absent	W < 5 m or W > 100 m	5 m ≤ W < 25 m	25 m ≤ W < 50 m	50 m ≤ W ≤ 100 m
6		Colour	Absent	Dark	Dark tan	Light tan/bleached	White/gold
7	Rocky shore	Slope	Absent	< 5°	5° – 10°	10° – 20°	> 20°
8		Extent	Absent	< 5 m	5 m – 10 m	10 m – 20 m	> 20 m
9		Roughness	Absent	Distinctly jagged	Deeply pitted and/or irregular	Shallow pitted	Smooth
10	Dunes		Absent	Remnants	Fore-dune	Secondary ridge	Several
11	Valley		Absent	Dry	Stream (< 1 m)	Stream (1 m – 4 m)	> 4 m
12	Skyline landforms		Not visible	Flat	Undulating	Highly undulating	Mountainous
13	Tides		Macro (> 4 m)		Meso (2 m – 4 m)		Micro (< 2 m)
14	Coastal landscape features[b]		None	1 feature	2 features	3 features	>3 features
15	Vistas		Open on one side	Open on two sides		Open on three sides	Open on four sides
16	Water colour & clarity		Muddy Brown/grey	Milky blue/green; opaque	Green/grey blue	Clear blue/dark blue	Very clear turquoise
17	Vegetation cover		Bare (< 10% vegetation only)	Scrub/Garigue/grass (marram/ferns, bramble/meadow, etc)	Wetland/meadow	Coppices, maquis (mature trees bushes)	Variety of mature trees/mature natural cover
18	Vegetation debris		Continuous > 50 cm high	Full strand line	Single accumulation	Few scattered items	None

(continued)

Table 4.1 (continued)

N	Physical Parameters	Rating				
		1	2	3	4	5
	Human parameters					
19	Disturbance factor (noise)	Intolerable	Tolerable		Little	None
20	Litter	Continuous accumulations	Full strand line	Single accumulation	Few scattered items	Virtually absent
21	Sewage (discharge evidence)	Sewage evidence		Some sewage evidence		No evidence of sewage
22	Non-built environment[c]	None		Hedgerow/terracing/monoculture		Field mixed cultivation ± trees/natural
23	Built environment[d]	Heavy industry	Heavy tourism and/or urban	Light tourism and/or urban and/or sensitive industry	Sensitive tourism and/or urban	Historic and/or none
24	Access type	No buffer zone/heavy traffic	Buffer zone/light traffic		Parking lot visible from coastal area	Parking lot not visible from coastal area
25	Skyline	Very unattractive	Unattractive	Sensitively designed	Very sensitively designed	Natural/historic features
26	Utilities[e]	>3 utilities	3 utilities	2 utilities	1 utility	None

[a]Cliff special features: Indentation, banding, folding, scree, irregular profile, etc.

[b]Coastal landscape features: Peninsulas, rock ridges, irregular headlands, arches, windows, caves, waterfalls, deltas, lagoons, islands, stacks, estuaries, reefs, fauna, embayment, tombola, etc.

[c]Non-built environment: When there is no agricultural activity.
If the natural vegetation cover parameter (17) has scored a 5, then tick the 5 box here.
If the natural vegetation cover parameter (17) has scored 2, 3, 4 then tick the 3 box here.

[d]Built environment: Caravans will come under tourism, grading 2: Large intensive caravan site, grading 3: Light, but still intensive caravan sites, grading 4: Sensitively designed caravan sites

[e]Utilities: Power lines, pipelines, street lamps, groins, seawalls, revetments, etc.

Table 4.2 Public Preference Surveys for the Scenic Parameters (Turkey, UK, Croatia and Malta, Ergin et al. 2011)

Country	Number of inquiries		Total number of inquiries
	Appropriate	Excluded	
British Council report (Turkey + UK) (2002)	270	–	270
Turkey (2003)	86	7	93
Croatia (2003)	56	51	107
Malta (2004)	73	0	73
Total	**485**	**58**	**543**

of us according to this personal history and for each of us, it varies also with mood, with purpose, and with attentiveness. What is seen, what is studied, and the way it is shaped and built in the landscape is selected and structured by custom, culture, desire, and faith. Environmental planning and improvement are merely academic exercises, doomed to failure without a prior understanding of the bases of perception and behavior.'

By the same reasoning, public surveys were statistically examined from Turkey, Malta, UK and Croatia, indicated that there were no significant variations between cross-cultural preferences in reflecting high public awareness for the environmental issues. As seen in the last column of Table 4.3, the absence of litter, good water colour and clarity, the absence of noise, the absence of buildings and utilities, coastal landscape features are highly appreciated parameters of coastal scenery (Ergin et al. 2011, Williams and Khatabi 2015).

4.2.2   Coastal Scenic Evaluation System (CSES)

The Coastal Scenic Evaluation System (CSES) is based on evaluation of a coastal region on a five point scale and provides preferential degree of coastal scenic parameters. The evaluation set, "E", based on such preference states, is defined as:

$$E = \left(\text{not at all}, \text{low}, \text{neutral}, \text{medium}, \text{high} \right) \tag{4.1}$$

The CSES (Table 4.1) was established as a checklist for evaluation of a coastal area on a five point rating scale stated above, classified from 1 to 5, implying a degree of preferences on the coastal scenic parameters as in Eq. 4.1 (Ergin et al. 2004, 2011). Physical Parameters (1–18) and Human Parameters (19–26) are given in column 2 of Table 4.1.

Table 4.3 Overall Public Survey Results for Croatia, Malta, Turkey and UK (Ergin et al. 2011)

Number of people contributed to the inquiry is 485								
Parameters			Importance					
No	Characteristics and features		1	2	3	4	5	Top five
1	Cliff	Height	63	71	160	114	77	22
2		Slope	77	102	159	82	65	10
3		Special features	58	72	119	99	137	23
4	Beach face	Type	44	39	72	101	229	136
5		Width	40	46	110	143	146	39
6		Colour	56	69	130	119	111	13
7	Rocky shore	Slope	80	120	154	86	45	4
8		Extent	69	115	155	89	57	10
9		Roughness	66	88	121	102	108	27
10	Dunes		111	103	119	86	66	11
11	Valley		65	54	92	146	128	31
12	Landform		59	59	89	106	172	52
13	Tides		121	96	140	64	64	18
14	Coastal landscape features		10	14	53	120	288	162*
15	Vistas		22	33	117	142	171	41
16	Water colour and clarity		6	4	15	73	387	333*
17	Vegetation cover		20	37	53	136	239	142
18	Vegetation debris		68	62	104	86	165	48
19	Disturbance factor (noise)		8	13	33	116	315	238*
20	Litter		7	4	17	38	419	371*
21	Sewage (discharge evidence)		7	4	17	38	419	371*
22	Non-built environment		67	65	141	108	104	40
23	Built environment		17	33	74	123	238	127
24	Access type		39	53	82	106	205	105
25	Skyline		5	14	42	109	315	213*
26	Utilities		5	14	42	109	305	213*

* The most important parameters

Within the fuzzy logic approach, the scenic assessment factor (set F) is expressed in terms of physical parameters (set P) and human parameters (set H) and given in set notation as:

$$F = (\text{Physical}, \text{Human}) = (P, H) \qquad (4.2)$$

Where subsets of set P and set H are formed from the following listed basic characteristics:

$$P = \left(\begin{array}{l} \text{cliff, beach face, rocky shore platform, dunes, valley and river mouth,} \\ \text{landform, tides, coastal landscape features, vistas of far places,} \\ \text{water colour and clarity, natural vegetation cover, vegetation debris} \end{array} \right) \qquad (4.3)$$

$$H = \begin{pmatrix} \text{noise,litter,sewage,non} - \text{built environment,} \\ \text{built environment,access type,skyline,utilities} \end{pmatrix} \quad (4.4)$$

The characteristics of cliff, beach face and rocky shore in Eq. (4.3) are further considered with the following sub-characteristics:

$$\text{Cliff} = (\text{height,slope,special features}) \quad (4.5a)$$

$$\text{Beach face} = (\text{type,width,colour}) \quad (4.5b)$$

$$\text{Rocky shore platform} = (\text{slope,extent,roughness}) \quad (4.5c)$$

4.2.3   Assessment Methodology

For ease of computations mostly matrix notation and operations are used throughout assessments in the methodology.

For factor F given in Eq. (4.2), a **weight matrix** W_F is defined with elements as P_w and H_w that are individual importance measures of the physical and human parameters, respectively. On the basis of an equal weight or importance assumption as suggested by experts for each of the parameter, sets P and H, W_F are given as:

$$W_F = \begin{bmatrix} P_W & H_W \end{bmatrix} = \begin{bmatrix} 0.5 & 0.5 \end{bmatrix} \quad (4.6)$$

The elements of the weight matrix are non-negative values whose sum is unity.

The **grade matrix** G_P for the subsets of P is assumed with equal entries g_i (i = 1 to 18) called grades (weights), for its parameters considering cliff, beach face and rocky shore characteristics as single entities, thus;

$$G_P = \begin{bmatrix} \frac{1}{36} & \frac{1}{36} & \frac{1}{36} & \frac{1}{36} & \frac{1}{36} & \frac{1}{36} & \frac{1}{36} & \frac{1}{36} & \frac{1}{36} & \frac{1}{12} & \frac{1}{12} & \frac{1}{12} & \frac{1}{12} & \frac{1}{12} & \frac{1}{12} & \frac{1}{12} & \frac{1}{12} & \frac{1}{12} \end{bmatrix} \quad (4.7)$$

Where the first nine weights correspond to cliff (height, slope, special features), beach face (type, width, colour) and rocky shore (slope, extent, roughness) properties, respectively. The last nine weights are for the parameters from dunes to vegetation debris.

Similarly, the **grade matrix** G_H for the subsets of H is assumed with equal entries with weights, g_i (i = 19 to 26) for its eight parameters from noise to utilities as:

$$G_H = \begin{bmatrix} \frac{1}{8} & \frac{1}{8} & \frac{1}{8} & \frac{1}{8} & \frac{1}{8} & \frac{1}{8} & \frac{1}{8} & \frac{1}{8} \end{bmatrix} \quad (4.8)$$

Grades (g_i) of the coastal scenic parameters demonstrate the significance of the feature compared to the others in the category (physical or human). The cliff, beach and rocky shore platform each had three sub-parameters. In order to give physical parameters the same significance, sub-parameters had ($g_i = 1/36$) grades and the other parameters in the physical category have ($g_i = 1/12$) grades, (Eq. 4.7). Likewise, human parameters, eight in total, had a significance grade of ($g_i = 1/8$), (Eq. 4.8).

4.2.4 Fuzzy Assessment

A fuzzy logic approach enables an expert group to quantify uncertainties and subjective pronouncements inherent in most scientific studies such as coastal scenic assessment evaluation (Zadeh 1965).

In CSES, by using a fuzzy mathematics approach, the membership functions, graphs, or matrices are defined for the representations of the magnitudes or weights of the inputs, relative to one another. The weights of the coastal scenic parameters were considered as fuzzy quantities and determined subjectively. They are a means of applying fuzzy logic rules to the inputs and for making fuzzy inferences about the outputs of the problem.

4.2.5 Weights of Coastal Scenic Parameters

To re-evaluate the validity of 'equal weights' of the coastal scenic parameters assumption and to bring out viewers' preferences and priority to different assessment parameters, the grades given in G_P and G_H were modified by the perception questionnaire results of public surveys (Table 4.3). Weight calculations for physical and human parameters are presented in (Tables 4.4 and 4.5), respectively. In calculation of the weights of parameters the following steps were carried out:

The **"Overall Weighted Averages w_i"** of physical and human parameters w_i were calculated for the voted attributes (given in the fourth and fifth boxes of Table 4.3 as given in Eq. 4.9) to promote higher preferences:

$$w_i = \frac{4N_{i,4} + 5N_{i,5}}{\left(number\ of\ public\ surveys\right)},\ i = 1\,to\,26 \tag{4.9}$$

Where $N_{i,4}$ and $N_{i,5}$ are the number of total scores ticked for the attributes 4 and 5 considered as in Box 4 and Box 5 (Table 4.3), respectively, for the ith parameter. For 485 public surveys, weighted averages (w_i) are given for all parameters in Tables 4.4 and 4.5.

- The **"Weights of Parameters $W_{P,i}$ and $W_{H,i}$"** for physical and human parameters with the ith entry were obtained, respectively as:

Table 4.4 Weights of physical parameters (Ergin et al. 2011)

i	Name	N_4 Box 4	N_5 Box 5	$w_{i=}(4N_{i,4}+5N_{i,5})/485$ Eq.(4.9)	$W_{Pi=}\ w_i \circ g_i$ Eq. (4.10a)	$W^*_{P,i}$ Eq. (4.11a)
	Physical parameters	Number of ticks (Table 4.3)				
1	Cliff height	114	77	1.7340	0.0482	0.0186
2	Cliff slope	82	65	1.3464	0.0374	0.0144
3	Cliff special features	99	137	2.2289	0.0619	0.0239
4	Beach type	101	229	3.1938	0.0887	0.0342
5	Beach width	143	146	2.6845	0.0746	0.0287
6	Beach colour	119	111	2.1258	0.0590	0.0227
7	Rocky shore slope	86	45	1.1732	0.0326	0.0126
8	Rocky shore extent	89	57	1.3216	0.0367	0.0141
9	Rocky shore roughness	102	108	1.9546	0.0543	0.0209
10	Dunes	86	66	1.3897	0.1158	0.0446
11	Valley	146	128	2.5237	0.2103	0.0810
12	Landform	106	172	2.6474	0.2206	0.0850
13	Tides	64	64	1.1876	0.0990	0.0381
14	Coastal landscape features	120	288	3.9588	0.3299	0.1271
15	Vistas	142	171	2.9340	0.2445	0.0942
16	Water colour and clarity	73	387	4.5918	0.3827	0.1474
17	Vegetation cover	136	239	3.5856	0.2988	0.1151
18	Vegetation debris	86	165	2.4103	0.2009	0.0774
				Total	2.5958	1.0000

$$W_{P,i} = w_i \circ g_i \ , \ i = 1\,to\,18 \tag{4.10a}$$

$$W_{H,i} = w_i \circ g_i \ , \ i = 19\,to\,26 \tag{4.10b}$$

Where ($\circ$) is an operator (inner product). g_i values were taken from Eqs. 4.7 and 4.8, respectively. $W_{P,i}$ and $W_{H,i}$ are given in columns 6 of Tables 4.4 and 4.5, respectively.

- The **"Normalized Final Weights W^*_p and W^*_H"** of the physical and human parameters separately obtained from the normalized forms of $W_{P,i}$ and $W_{H,i}$ with the i'th entry are defined respectively as:

Table 4.5 Weights of human parameters (Ergin et al. 2011)

I	Name	Number of ticks		$w_i = (4N_{i,4} + 5 N_{i,5})/485$ Eq. (4.9)	$W_{H,i} = w_i \circ g_i$ Eq. (4.10a)	$W^*_{H,i}$ Eq. (4.11b)
		N4 Box 4	N5 Box 5			
19	Disturbance factor (noise)	116	315	4.2041	0.5255	0.1362
20	Litter	38	419	4.6330	0.5791	0.1501
21	Sewage (discharge evidence)	38	419	4.6330	0.5791	0.1501
22	Non-built environment	108	104	1.9629	0.2454	0.0636
23	Built environment	109	315	4.1464	0.5183	0.1344
24	Access type	106	205	2.9876	0.3735	0.0968
25	Skyline	109	315	4.1464	0.5183	0.1344
26	Utilities	109	315	4.1464	0.5183	0.1344
				Total	3.8575	1.0000

$$W^*_{P,i} = \frac{w_i \circ g_i}{\sum_{18}^{i=1} w_i \circ g_i} = \frac{W_{P,i}}{\sum_{18}^{i=1} w_i \circ g_i}, \; i = 1\,to\,18 \qquad (4.11a)$$

$$W^*_{H,i} = \frac{w_i \circ g_i}{\sum_{18}^{i=1} w_i \circ g_i} = \frac{W_{H,i}}{\sum_{18}^{i=1} w_i \circ g_i}, \; i = 19\,to\,26 \qquad (4.11b)$$

Normalized final weights of parameters W^*_P and W^*_H (physical and human respectively) to be used in the fuzzy logic applications are presented in the last columns of Tables (4.4 and 4.5).

- The **"Final Weight Matrix W*"** was formed from the normalized weight parameters W^*_P and W^*_H as:

$$W^* = \begin{bmatrix} W^*_P & W^*_H \end{bmatrix} \qquad (4.12)$$

4.2.6 Fuzzy Logic Matrices

The evaluation of a coastal area by experts and trained groups depends on education, age, gender, national and cultural background, and even the training of the surveyors. During the evaluation process, factors such as time constraints (weather, seasonal or temporary conditions) and biases of the group may be effective.

Therefore the evaluation of a coastal area by an expert might be highly subjective, uncertain and fuzzy (see Chap. 2).

Within this study the aim was to identify and assess the highest coastal scenic values possible to be used as tool for management and conservation of scenic quality. For all coastal sites, scenic assessments were carried out during fine weather and seasonal conditions, to have a similar comparative basis. Similarly, the space span of the coastal site was decided by considering the best scenic parameters within the study area in order to enhance landscape preservation and protection.

The **"Membership-grade matrices, M_i (i = 1 to 26)"** for all parameters are the basis for analysis of an attribute rating system by a fuzzy logic approach. Each M_i for the ith scenic parameter was developed by considering degree of variation or error when the expert is obliged to make a single decision among several other possible grades over an attribute. To overcome subjectivity and bias in the assessment process and to quantify uncertainty, a fuzzy logic mathematical methodology was adopted to this checklist approach. Membership-grade matrix abates Boulean logic in the input matrices and reverberates the fuzziness in the methodology (Zadeh 1965).

In CSES, the membership-grade matrices, (M_i) represent different membership measures among neighbouring attributes with grading in the interval [0, 1]. The membership-grade matrix (M_i) of the ith parameter was formed as a square matrix of order 5, where the elements show the degree of the relationship of the voted attributes to the remaining attributes. The elements of (M_i) were formed from possibilities ranging from 0 to 1, where 0 implies no possibility and 1 implies the highest possibility on the attributes (BCR 2003).

If the parameter was absent or not relevant then the first element of the first row is 1, while all other entries of this row are 0 denoting the absoluteness of the grade 'absent' or 'not relevant'.

If the first element of the first row is not 1 then the idea of error was introduced on the chosen grades when one is obliged to make a unique decision among several other possible grades over the attribute. In this case the possibility of existence of other attributes was shared between the neighbouring attributes with possibility measures ranging between larger than 0 and less than 1. Values for the possibilities of errors were based on the studies of expert opinions (BCR 2003).

The membership-grade matrices (M_i) of the 26 scenic parameters are presented in Appendix 2.

4.2.7   An Example for a Coastal Scenic Evaluation System

The basic steps of coastal site scenic evaluations are presented as an example for Çıralı, Karaburun, Turkey (BCR 2003) and given in "Coastal Scenic Evaluation System" (Table 4.6) where the columns C1 to C19 are as follows:

- Column C1: is for parameter number (1 to 26).
- Column C2: is for scenic parameters (physical and human).

Table 4.6 Input Matrices for Çıralı Karaburun, Turkey (BCR 2004)

No	Assessment parameters	Weights of parameters	Graded attributes	Input matrices (di)					Gi Matrices	Grade matrices (Gi) (attributes, 1–5)					Ri Matrices	Fuzzy assessment matrix (attributes, 1–5)				
C1	C2	C3	C4	C5 to C9						C10 to C14						C15 to C19				
	Physical																			
1	Cliff height	0.019	3	0	0	1	0	0	G_P	0.00	0.30	1.00	0.30	0.00	R_P	0.000	0.006	0.019	0.006	0.000
2	Cliff slope	0.014	5	0	0	0	0	1		0.00	0.00	0.00	0.50	1.00		0.000	0.000	0.000	0.007	0.014
3	Special features	0.024	3	0	0	1	0	0		0.00	0.00	1.00	0.30	0.00		0.000	0.000	0.024	0.007	0.000
4	Beach type	0.034	4	0	0	0	1	0		0.00	0.00	0.00	1.00	0.00		0.000	0.000	0.000	0.034	0.000
5	Beach width	0.029	4	0	0	0	1	0		0.00	0.00	0.20	1.00	0.60		0.000	0.000	0.006	0.029	0.017
6	Beach color	0.023	4	0	0	0	1	0		0.00	0.00	0.60	1.00	0.00		0.000	0.000	0.014	0.023	0.000
7	Shore slope	0.013	1	1	0	0	0	0		1.00	0.00	0.00	0.00	0.00		0.013	0.000	0.000	0.000	0.000
8	Shore extent	0.014	1	1	0	0	0	0		1.00	0.00	0.00	0.00	0.00		0.014	0.000	0.000	0.000	0.000
9	Shore roughness	0.021	1	1	0	0	0	0		1.00	0.00	0.00	0.00	0.00		0.021	0.000	0.000	0.000	0.000
10	Dunes	0.045	2	0	1	0	0	0		0.00	1.00	0.00	0.00	0.00		0.000	0.045	0.000	0.000	0.000
11	Valley	0.081	4	0	0	0	1	0		0.00	0.00	1.00	1.00	0.10		0.000	0.000	0.000	0.081	0.008
12	Landform	0.085	5	0	0	0	0	1		0.00	0.00	0.00	0.20	1.00		0.000	0.000	0.000	0.017	0.085
13	Tides	0.038	5	0	0	0	0	1		0.00	0.00	0.00	0.00	1.00		0.000	0.000	0.000	0.000	0.038
14	Landscape features	0.127	3	0	0	1	0	0		0.00	1.00	1.00	0.20	0.00		0.000	0.127	0.127	0.025	0.000
15	Vistas	0.094	4	0	0	0	1	0		0.00	0.00	0.00	1.00	0.30		0.000	0.000	0.000	0.094	0.028
16	Water color	0.147	5	0	0	0	0	1		0.00	0.00	0.00	0.20	1.00		0.000	0.000	0.000	0.029	0.147
17	Vegetation cover	0.115	5	0	0	0	0	1		0.00	0.00	0.00	0.20	1.00		0.000	0.000	0.000	0.023	0.115
18	Vegetation debris	0.077	5	0	0	0	0	1		0.00	0.00	0.00	0.20	1.00		0.000	0.000	0.000	0.015	0.077

Weighted averages matrix for subset physical V_P

	Human			G_H					R_H						0.048	0.051	0.190	0.390	0.529
19	Disturbance factor (noise)	0.136	5	0	0	0	0	1	0.00	0.00	0.00	0.20	1.00		0.000	0.000	0.000	0.027	0.136
20	Litter	0.150	4	0	0	0	1	0	0.00	0.00	0.20	1.00	0.20		0.000	0.000	0.030	0.150	0.030
21	Sewage	0.150	5	0	0	0	0	1	0.00	0.00	0.20	0.00	1.00		0.000	0.000	0.030	0.000	0.150
22	Non-built environment	0.064	5	0	0	0	0	1	0.00	0.00	0.20	0.00	1.00		0.000	0.000	0.013	0.000	0.064
23	Built environment	0.134	4	0	0	0	1	0	0.00	0.00	0.30	1.00	0.00		0.000	0.000	0.040	0.134	0.000
24	Access type	0.097	4	0	0	0	1	0	0.00	0.20	0.00	1.00	0.20		0.000	0.019	0.000	0.097	0.019
25	Skyline	0.134	5	0	0	0	0	1	0.00	0.00	0.00	0.00	1.00		0.000	0.000	0.000	0.000	0.134
26	Utilities	0.134	5	0	0	0	0	1	0.00	0.00	0.00	0.20	1.00		0.000	0.000	0.000	0.027	0.134
	Weighted averages matrix for subset human V_H														**0.000**	**0.019**	**0.113**	**0.435**	**0.667**

Weighted averages matrix

Elements of weighted averages matrix	Weights of subsets W_F	Matrix K	Attributes (1–5)				
			1	2	3	4	5
Weighted averages matrix of subset physical V_P	1/2		0.048	0.051	0.190	0.390	0.529
Weighted averages matrix of subset human V_F	1/2		0.000	0.019	0.113	0.435	0.667

Final assessment matrix $R = (W_F K)$

Final assessment matrix (R) – membership degree			0.024	0.035	0.152	0.413	0.598

- Column C3 is for the weights of parameters (taken from Tables 4.4 and 4.5).
- Column C4: is for graded attributes from the scenic assessment.
- Columns C5 to C9: are for the "**Input Matrix (d_i)**" of order 1x5 for each scenic parameter, with corresponding attributes from 1 to 5 (the value was given 1 for the attribute scored and 0 for the other attributes).

For this site, if the graded attribute for 'beach face width (i = 5)', is 4, then input matrix $d_i = d_5$ is

$$d_5 = \begin{bmatrix} 0 & 0 & 0 & 1 & 0 \end{bmatrix} \tag{4.13}$$

- Columns C10 to C14: are for the "Grade Matrices G_i (i =19 to 26)".

The grade matrix G_i of the ith parameter is defined as:

$$G_i = d_i \times M_i \tag{4.14}$$

Where ($\times$) operator stands for matrix multiplication. For the parameter 'beach face width (W); i = 5' the corresponding membership grading matrix M_5 (Appendix 2) is

$$M_5 = \begin{matrix} & 1 & 2 & 3 & 4 & 5 \\ \begin{bmatrix} 1 & 0 & 0 & 0 & 0 \\ 0 & 1 & 0 & 0 & 0 \\ 0 & 0.2 & 1 & 0.2 & 0 \\ 0 & 0 & 0.2 & 1 & 0.6 \\ 0 & 0 & 0 & 0.6 & 1 \end{bmatrix} & \begin{matrix} 1: \textit{Stands for "Absent"} \\ 2: W\langle 5\text{m } \textit{or } W\rangle 100\,\text{m} \\ 3: 5\text{m} \leq W < 25\,\text{m} \\ 4: 25\text{m} \leq W < 50 \text{ m} \\ 5: 50 \text{ m} \leq W \leq 100\,\text{m} \end{matrix} \end{matrix}$$

In the above matrix M_5, first two rows show that if attributes 1 and 2 are ticked then it is extremely likely that other attributes are not possible. Similarly, if it scores 3 the error could be on either side of the true grade, and a possible error of 0.2 is given on either side. In the same way if it scores four or five the error is distributed according to a logical point of view, a higher logical possibility implies higher error distribution.

Accordingly, the grade matrix of 'Beach Face Width' is:

$$G_5 = d_5 \times M_5 = \begin{bmatrix} 0 & 0 & 0 & 1 & 0 \end{bmatrix} \times \begin{bmatrix} 1 & 0 & 0 & 0 & 0 \\ 0 & 1 & 0 & 0 & 0 \\ 0 & 0.2 & 1 & 0.2 & 1 \\ 0 & 0 & 0.2 & 1 & 0.6 \\ 0 & 0 & 0 & 0.6 & 1 \end{bmatrix} \tag{4.15}$$

$$= \begin{bmatrix} 0 & 0 & 0.2 & 1 & 0.6 \end{bmatrix}$$

The grade matrices (G_i) are given in Table 4.6 in columns C10 to C14.

The matrices A_P and A_H for the scenic parameters in the sets P and H, are formed separately for each of the factors P and H, respectively. A_P is an (18×5) matrix formed from grade matrices G_i (i = 1 to 18) as its rows. Similarly, A_H is a (6×5) matrix formed from grade matrices G_i (i = 19 to 26) as its rows and presented in columns C10 to C14.

- Columns C15 to C19: are for the **"Fuzzy Assessment Matrices R_P and R_H."**

R_P and R_H are formed separately for the scenic parameters in the sets P and H respectively as presented in Table 4.6. R_P is an (18×5) matrix formed from (i = 1 to 18) rows and similarly R_H is a matrix (6×5) formed with (i = 19 to 26) as its rows, respectively:

$$R_{P,i} = W_{P,i}{}^* . G_i, \ i = 1 \, to \, 18 \tag{4.16}$$

$$R_{H,i} = W_{H,i}{}^* . G_i, \ i = 19 \, to \, 26 \tag{4.17}$$

Rounding W* values to three decimal places for i = 5, $R_{P,5}$ becomes

$$\begin{aligned} R_{P,5} &= 0.029. \begin{bmatrix} 0 & 0 & 0.2 & 1 & 0.6 \end{bmatrix} \\ &= \begin{bmatrix} 0 & 0 & 0.006 & 0.029 & 0.017 \end{bmatrix} \end{aligned} \tag{4.18}$$

The fuzzy assessment matrices R_P and R_H are presented in Table 4.6 in columns C15 to C19.

- The **"Weighted Average Matrices V_P and V_H"** are the column summation of Fuzzy Assessment Matrices R_p and R_H for the physical and human subsets respectively and are given in Eqs. 4.19a and 4.19b as:

$$V_P = \begin{bmatrix} 0.048 & 0.051 & 0.190 & 0.390 & 0.529 \end{bmatrix} \tag{4.19a}$$

$$V_H = \begin{bmatrix} 0.000 & 0.019 & 0.113 & 0.435 & 0.667 \end{bmatrix} \tag{4.19b}$$

The Weighted Average Matrices V_P and V_H are given in Table 4.6.

- The **"Weighted Average Matrix K"** is a 2×5 matrix formed from V_P and V_H as its rows, respectively;

$$K = \begin{bmatrix} V_P \\ V_H \end{bmatrix} \tag{4.20}$$

$$K = \begin{bmatrix} 0.048 & 0.051 & 0.190 & 0.390 & 0.529 \\ 0.000 & 0.019 & 0.113 & 0.435 & 0.667 \end{bmatrix}$$

Matrix K is also presented in Table 4.6.

- The "**Final Assessment Matrix R** (named **Membership Degree**)" is a matrix of 1×5 obtained by multiplying K with subsets weights as given in Table 4.6.

$$R = W_F \times K \qquad (4.21)$$

$$R = \begin{bmatrix} 0.5 & 0.5 \end{bmatrix} \times \begin{bmatrix} 0.048 & 0.051 & 0.190 & 0.390 & 0.529 \\ 0.000 & 0.019 & 0.113 & 0.435 & 0.667 \end{bmatrix} \qquad (4.22)$$

$$R = \begin{bmatrix} 0.024 & 0.035 & 0.152 & 0.413 & 0.598 \end{bmatrix} \qquad (4.23)$$

The Final Assessment Matrix R is presented in Table 4.6.

4.2.7.1 Data Presentation

Coastal Scenic Assessment data for Çıralı Karaburun is presented in scenic assessment and weighted average histogram of physical and human parameters (Fig. 4.1) weighted average histogram (Fig. 4.2) together with membership degree graph (Fig. 4.3).

Scenic Assessment Histogram This was obtained by plotting the scores taken from the 'Coastal Scenic Evaluation System' (Table 4.6) on the y-axis for each parameter versus the scenic evaluation ratings on the x axis. The x-axis is further grouped into physical and human subsections. The scenic assessment histogram of Çıralı Karaburun for physical and human parameters is presented in Fig. 4.1. The histogram obtained shows higher ratings for human parameters having almost uniform distribution among themselves (implying a fine-toothed comb model with all ratings above 4) than the physical ones. The high scores for human parameters show the positive effects of appropriate utilization of the coastal site. The opposite result is true for physical parameters with ratings ranging from 1 to 5.

Weighted Averages Histograms of Physical and Human Parameters (V_P, V_H) The histogram of the weighted averages grouped into physical (V_P) and human (V_H) parameters versus the attributes from 1 to 5 are given in Fig. 4.2.

Membership Degree Graphs (Final Assessment Matrix R) The graph of the membership degrees of physical and human parameters versus the attributes from 1 to 5 (Final Assessment Matrix R) is presented in Fig. 4.3.

4.2.8 Data Interpretation and Comparative Site Studies

For a comparative site assessment study, Kızkalesi Mersin and Çıralı Karaburun are chosen where the first site is as an example of overbuilding and extensive coastal development of a highly urbanized coastal site and the latter is an example of an appropriate coastal utilization.

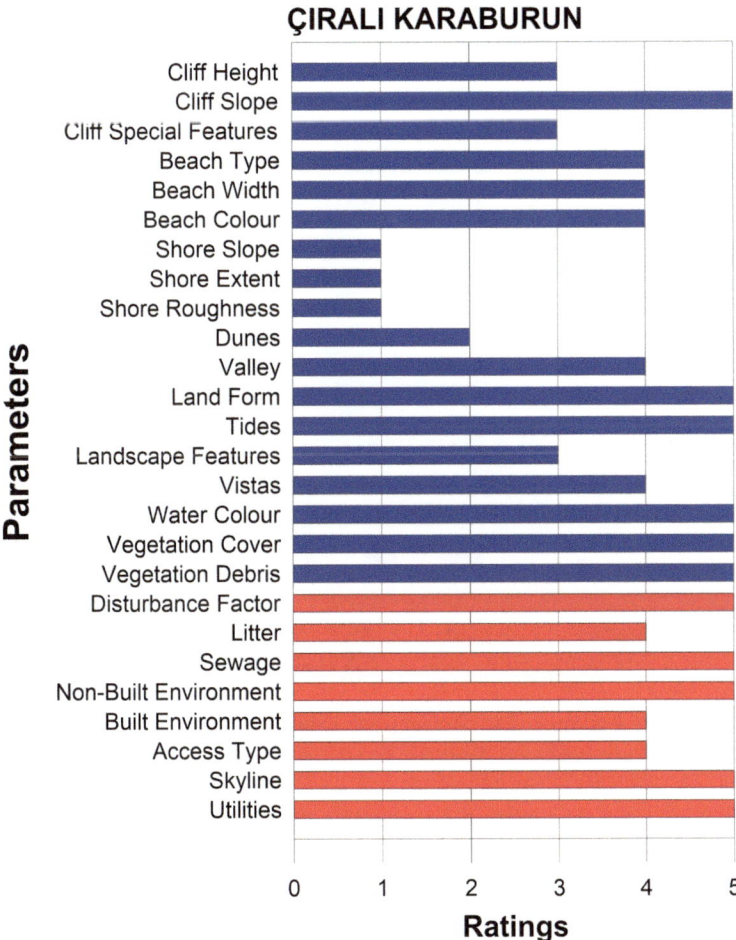

Fig. 4.1 Assessment histogram for Çıralı Karaburun, Turkey (BCR, 2004)

The scenic assessment histogram of Kızkalesi Mersin for physical and human parameters is presented in Fig. 4.4. The histogram obtained shows lower ratings for human parameters with almost all ratings below 3. The low grades for human parameters indicate the negative effects of human utilization of the coastal site, such as, unattractive urbanization, intensive development.

The weighted average histogram and membership degree versus attributes of Kızkalesi Mersin are given in Figs. 4.5 and 4.6, respectively.

With respect to the weighted averages versus attributes histogram for Çıralı Karaburun (Fig. 4.2), high weighted averages at high attribute values such as 4 and 5 reflect the positive influencing impact of the physical and/or human parameters. The reverse holds true for lower attribute values such as 1 and 2, which

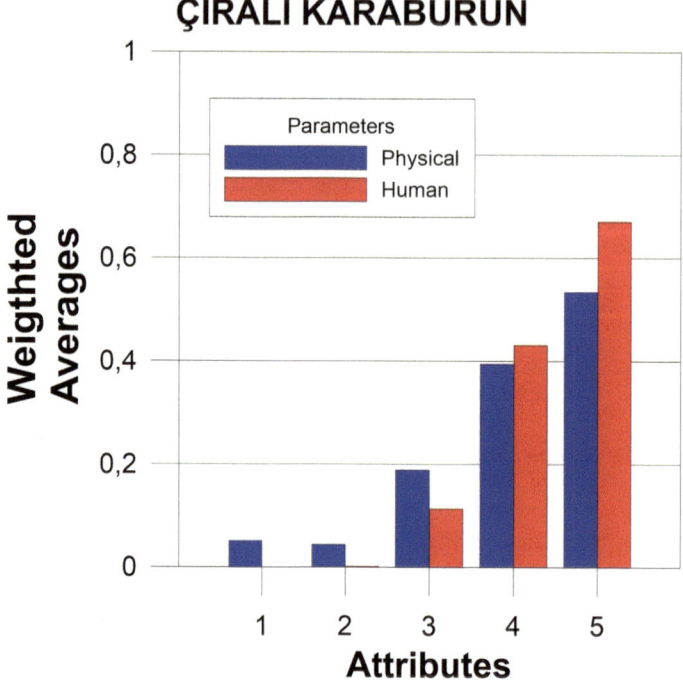

Fig. 4.2 Weighted Averages *vs* Attributes for Çıralı Karaburun, Turkey (BCR 2004)

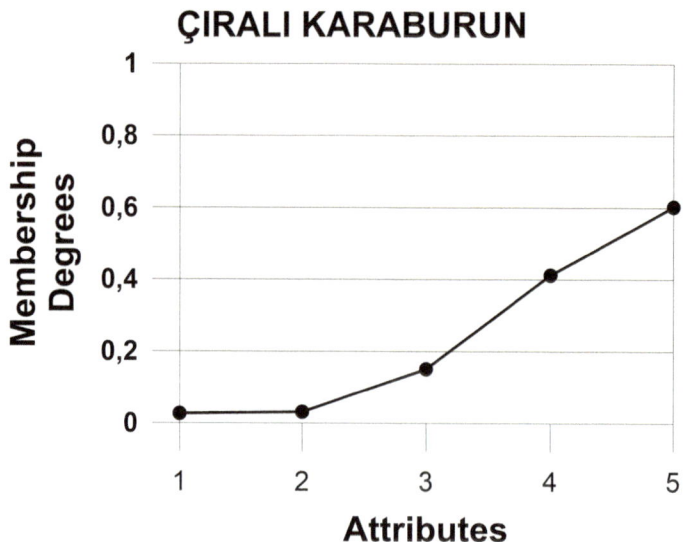

Fig. 4.3 Membership Degrees *vs* Attributes for Çıralı Karaburun, Turkey (BCR 2004)

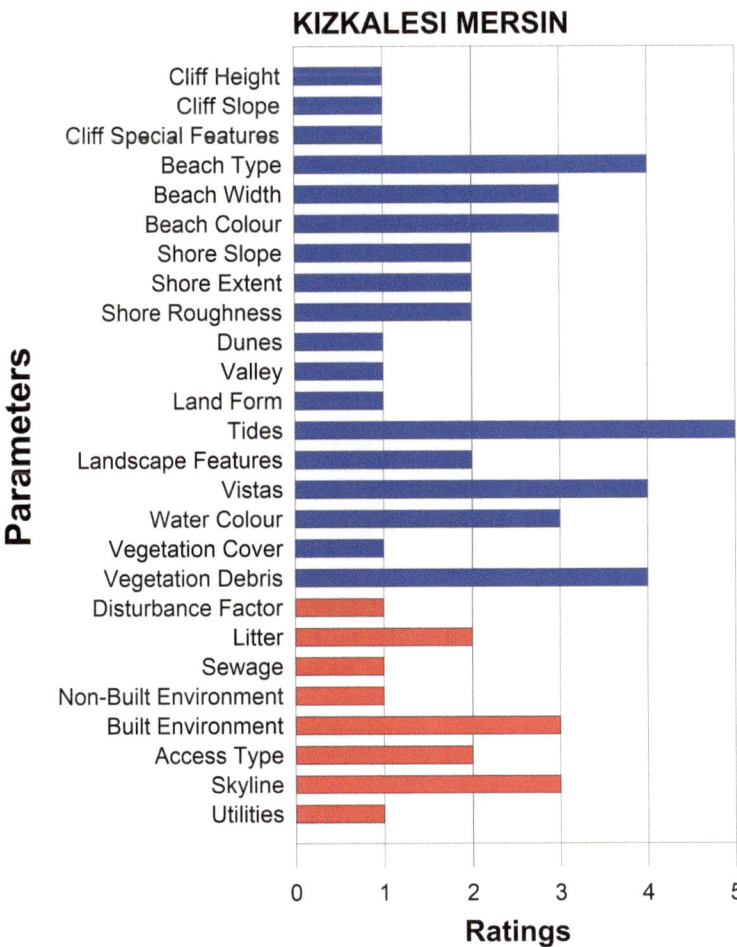

Fig. 4.4 Scenic Assessment histogram for Kızkalesi Mersin, Turkey (BCR 2004)

reflect the adverse impact of the physical and/or human parameters at Kızkalesi Mersin (Fig. 4.5). With respect to membership degree graphs at Çıralı Karaburun (Fig. 4.3), a right-hand skew indicates a high scenic rating, compared to a low scenic one, epitomised by a left-hand skew curve as presented in Fig. 4.6 for Mersin Kızkalesi.

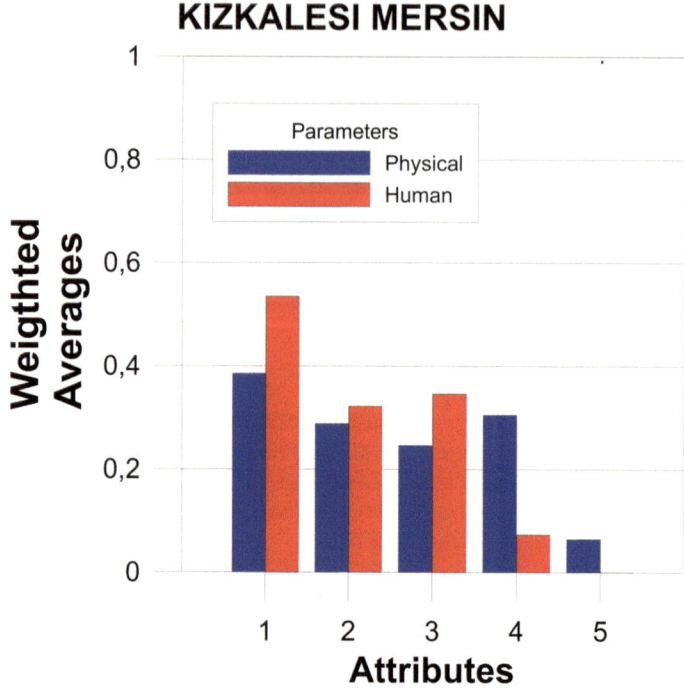

Fig. 4.5 Weighted Averages *vs* Attributes for Kızkalesi Mersin, Turkey (BCR 2004)

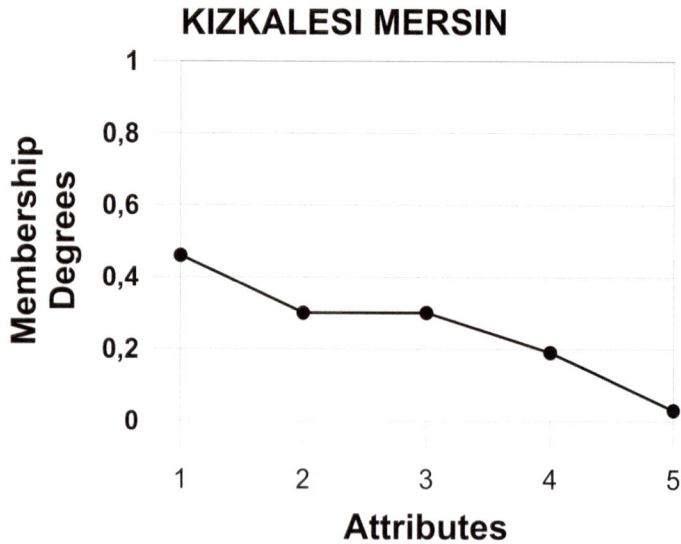

Fig. 4.6 Membership Degrees *vs* Attributes for Kızkalesi Mersin Turkey (BCR 2004)

4.2.9   Decision Parameters

To compare the sites, a special 'decision parameter' which is referred to as evaluation index (D) was defined using the 'Membership Degrees' graphs (Ergin et al. 2004). In these graphs, in principle, the maximum membership degree permits decisions on the state that corresponds to the entry with the maximum value and disregards the choices made over the other states of the set E of Eq. (4.1). The evaluation index (D) was introduced as further scoring to take into consideration differences among the states of the set E where the uncertainty is distributed via membership grade matrices.

D values were estimated as the ratio of the area A_{ij} between the ith and jth attribute to that of the selected area under the 'Membership Degree Curve'. A set arithmetically defined decision parameters (D1 – D4) were suggested as follows;

$$D1 = \frac{A_{35}}{A_T} \tag{4.24}$$

$$D2 = \frac{A_{35}}{A_{13}} \tag{4.25}$$

$$D3 = \frac{A_{35} - A_{13}}{A_T} \tag{4.26}$$

$$D4 = \frac{(-2A_{12}) + (-A_{23}) + (A_{34}) + (2A_{45})}{A_T} \tag{4.27}$$

In Eqs. 4.24, 4.25, 4.26 and 4.27:

- A_T is the total area under the attribute curve and the area under the curve between attributes 1 and 2 is named A_{12}.
- the area under the curve between the attributes 2 and 3 is named A_{23}.
- the area under the curve between the attributes 3 and 4 is named A_{34}.
- the area under the curve between the attributes 4 and 5 is named A_{45}.

Whereas the area under the curve between attributes 1 and 3, i.e. "$A_{12} + A_{23}$" is named A_{13}; the area under the curve between the attributes 3 and 5, i.e. "$A_{34} + A_{45}$" is named A_{35}. The total area under the curve is denoted by $A_T = A_{12} + A_{23} + A_{34} + A_{45}$.

The above calculations were carried out for all evaluated sites using decision parameters, D1 to D4. The membership degree graph for Çıralı Karaburun (Fig. 4.3) is used as an example to compute decision parameters D1 to D4. The areas for the D value calculations for Çıralı Karaburun are given in Table 4.7.

D1 to D4 values for Çıralı Karaburun and Kızkalesi Mersin were obtained and listed in Table 4.8:

Table 4.7 Area Calculations from Membership Degree Graphs for Çıralı Karaburun, Turkey (BCR 2004)

Areas below the curve for all attributes

1–2		2–3		3–4		4–5			Attribute>3 (3–5)		Attribute<3 (1–3)		
A_{12}	A_{12}/A_T	A_{23}	A_{23}/A_T	A_{34}	A_{34}/A_T	A_{45}	A_{45}/A_T	A_T	A_{35}	A_{35}/A_T	A_{13}	A_{13}/A_T	A_{35}/A_{13}
0.029	0.032	0.093	0.102	0.282	0.310	0.506	0.556	0.911	0.789	0.866	0.122	0.134	6.455

Table 4.8 D Values obtained for Çıralı Karaburun and Kızkalesi Mersin Turkey (BCR 2004)

Site	D1	D2	D3	D4
Çıralı, Karaburun	0.87	6.46	0.73	1.26
Kızkalesi Mersin	0.34	0.52	−0.32	−0.59

For a given site, all D1 to D4 values reflected almost similar assessment results. D4 (Eq. 4.27) was selected as the 'Evaluation Index (D)' as a decision tool since it reflects the attribute values in terms of weighted areas. Evaluation Index (D4), Eq. (4.27), gives the results biased for higher attribute values (e.g. area between 4 and 5 is multiplied by 2) to enhance the protection of sites in Classes 4 and 5. Similarly, the results obtained for lower attributes are biased for lower attributes (e.g. area between 1 and 2 is multiplied by −2) to stress the adverse results of an intensive development. Since class grading is also linguistic and no definite boundaries exist between the classes, D value aims and considers relative measures in membership grade matrix elements. D4 value, as defined in (Eq. 4.27) changes from (−2 to +2).

Within the British Council Project, pilot studies (from 2001 to 2004) coastal scenic evaluation assessment of some coastal sites were carried out in 57 coastal sites of Turkey, UK and Malta together with 33 sites from Croatia along the Dalmatian coastline. Site assessments and related studies have been presented in several reports and papers (BCR 2003, BCR 2004 and Ergin et al. 2004).

The Evaluation Index (D) values of these coastal areas from Turkey (22 sites), UK (18 sites), Malta (17 sites) and Croatia (33 sites) are presented in Table 4.9 together with a graphical form in Fig. 4.7. A plot of the evaluation index D for 90 sites is also presented in Fig. 4.8 with cumulative relative frequencies versus Z-score D values. The normalized D-values and their cumulative distribution (Eqs. 4.28 and 4.29) for 90 sites showing a Gaussian distribution (normal distribution) were a classification base for the other assessed coastal sites provided that they are assessed under the same evaluation parameters used in CSES.

In the last decade, the authors of this book have assessed 952 global coastal sites by implanting CSES. Some results will be presented in Chap. 6.

Normal plotting is a graphical method for determining whether sample data conform to normal distribution based on a subjective visual examination of the data.

$$Z-score = \frac{D-\left(Average \ of \ D\right)}{Standard \ Deviation \ of \ D} \tag{4.28}$$

$$Cumulative \ Relative \ Frequency : F_D\left(d\right) = \frac{n-k+0.5}{n} \tag{4.29}$$

Where k is the number showing the coastal site listed with D values from maximum to minimum as given n Table 4.9 and n is the total number of sites.

Table 4.9 D Values of 90 Coastal Sites from Turkey (TR), United Kingdom (UK), Malta (MT) and Croatia (CR). (BCR 2004 and Ergin et al. 2004)

N	Site	D	Class	N	Site	D	Class
1	Çıralı mid-section (TR)	1.31	I	46	Sustipan (CR)	0.44	III
2	Çıralı Karaburun (TR)	1.26	I	47	Ploce (CR)	0.40	III
3	Zlatni rat (CR)	1.21	I	48	Mellieha (MT)	0.37	IV
4	Zaglav (CR)	1.18	I	49	Wisemans bridge (UK)	0.34	IV
5	BigLake (CR)	1.09	I	50	Broadhaven (UK)	0.34	IV
6	Phasalis Small Bay (TR)	1.08	I	51	Angle (UK)	0.33	IV
7	Struga (CR)	1.08	I	52	Bonje (CR)	0.33	IV
8	Zabarje (CR)	1.07	I	53	Alata west, Mersin (TR)	0.31	IV
9	Sv. Miahljo (CR)	1.06	I	54	Alata mid, Mersin (TR)	0.29	IV
10	Little Haven (UK)	1.00	I	55	Tenby North (UK)	0.26	IV
11	Dingli cliffs (MT)	0.97	I	56	Bobovisca Luka (CR)	0.26	IV
12	Phaselis large bay (TR)	0.91	I	57	Albatross beach, Cav. (CR)	0.26	IV
13	Poppit (UK)	0.91	I	58	Dance (CR)	0.25	IV
14	Tisan back bay, Mersin(TR)	0.83	II	59	Kastelet (CR)	0.22	IV
15	Mezuporat (CR)	0.83	II	60	Milna (CR)	0.21	IV
16	Zaklopatica (CR)	0.82	II	61	Antalya Old Harbour(TR)	0.19	IV
17	Korcula (CR)	0.80	II	62	Tekirova north (TR)	0.19	IV
18	Polace (CR)	0.78	II	63	Tekirova south (TR)	0.18	IV
19	Fungus rock (MT)	0.77	II	64	Kercem cliffs (MT)	0.16	IV
20	Nash Point (UK)	0.74	II	65	Saundersfoot (UK)	0.15	IV
21	St Govans (UK)	0.69	II	66	Supetar (CR)	0.13	IV
22	Srebrna (CR)	0.69	II	67	Konyaaltı west (TR)	0.10	IV
23	Tisan Tample, Mersin (TR)	0.68	II	68	White towers (MT)	0.10	IV
24	Whitesands (UK)	0.68	II	69	Konyaaltı east (TR)	0.09	IV
25	Stiniva (CR)	0.68	II	70	Xwieni point (MT)	0.08	IV
26	Karaburun Akyar, Mersin (TR)	0.67	II	71	Xlendi Bay (MT)	0.07	IV
27	Salbunara (CR)	0.67	II	72	Alata east, Mersin (TR)	0.07	IV
28	Skrivena Luka (CR)	0.67	II	73	Llantwit Major (UK)	0.04	IV
29	Newgale (UK)	0.66	II	74	Konyaaltı middle (TR)	0.04	IV
30	Sobra (CR)	0.66	II	75	Ogmore (UK)	0.03	IV
31	StariGrad (CR)	0.66	II	76	Porthcawl (UK)	0.02	IV
32	Porat (CR)	0.65	II	77	Antalya waterfalls(TR)	−0.01	V
33	Göksu Hurma, Mersin(TR)	0.61	III	78	Mygarr ix-xini (MT)	−0.02	V
34	Tenby South (UK)	0.57	III	79	Ramla Bay (MT)	−0.06	V
35	Ghajn Tuffieha (MT)	0.56	III	80	Amroth (UK)	−0.08	V
36	Manikata (MT)	0.56	III	81	Ghallis rocks coastline (MT)	−0.12	V
37	Southerndown (UK)	0.54	III	82	Antalya Lara Barınak (TR)	−0.16	V
38	Komiza 2 (CR)	0.54	III	83	Antalya Dedeman Hotel (TR)	−0.21	V
39	Komiza 1 (CR)	0.51	III	84	Uvala Lapad (CR)	−0.26	V

(continued)

Table 4.9 (continued)

N	Site	D	Class	N	Site	D	Class
40	Milna (CR)	0.49	III	85	BaVice (CR)	−0.26	V
41	Calypso cave (MT)	0.48	III	86	Lara Beach (TR)	−0.28	V
42	FreshWater West (UK)	0.46	III	87	Marsalforn (MT)	−0.37	V
43	Croatia beach, Cavtat (CR)	0.46	III	88	Bahar Ic-caghaq (MT)	−0.41	V
44	Jadran (CR)	0.46	III	89	Kızkalesi, Mersin (TR)	−0.58	V
45	Blue lagoon (UK)	0.45	III	90	St. George's bay (MT)	−0.64	V

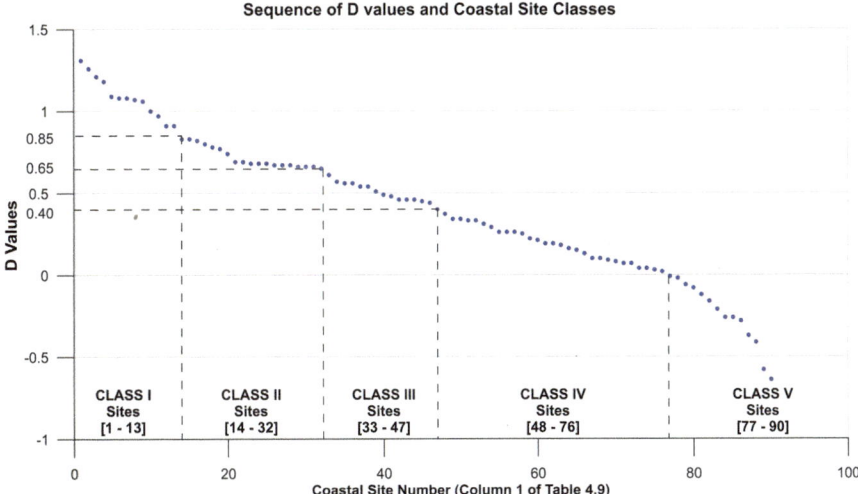

Fig. 4.7 Coastal Scenic Sequence Curve with respect to D Value of 90 Sites

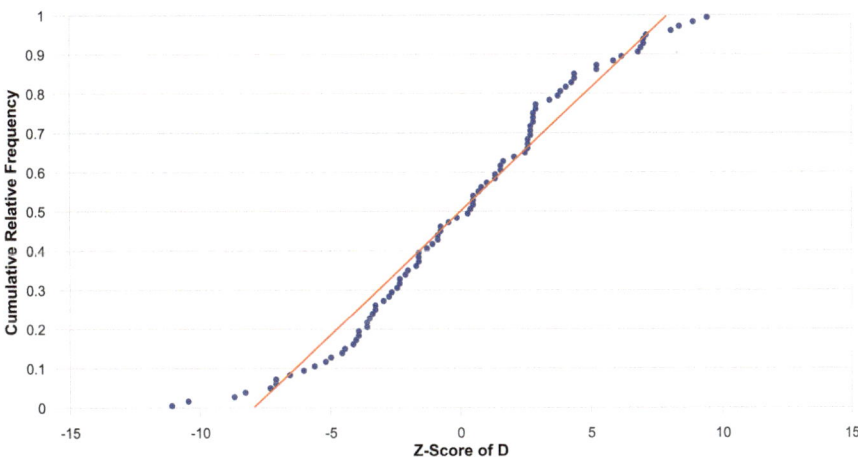

Fig. 4.8 Normal Plot of Evaluation Index (D) for 90 sites

Table 4.10 Classification of Coastal Areas According to Their D Values

Class	D values	Main features of the coastal sites
I	$D \geq 0.85$	**Top natural**: Extremely attractive sites with very high landscape value
II	$0.85 > D \geq 0.65$	**Natural**: Attractive sites with high landscape value
III	$0.65 > D \geq 0.40$	**Natural**: Average sites with medium landscape value (with the exception of Tenby south, which is an urban site with exceptional scenic characteristics)
IV	$0.40 > D \geq 0.00$	**Mainly urban**: Poor sites with medium landscape value and light development
V	$D < 0.00$	**Urban**: Poor sites with low landscape value and intensive development

4.3 Site Classifications

I must go down to the sea again. John Masefield, Sea Fever

Site classifications were made on the final sequence curve produced, based on the Evaluation Index D (Fig. 4.6). Curve break points based on the midpoints change of slope, allowed an obvious division of sites into five main classes. The break point statistical distribution, were expected to be Gaussian (normal) indicating unbiasedness. For this purpose normality tests using chi-square and Kolmogorov-Smirnov tests were performed to the 5% significance level conforming normality of the break point distributions (Ergin et al. 2004).

In essence, the higher the D value a coastal site has, the better is the scenic value. The sites with negative D values are considered the poorest and are referred to as Class V in the study. The limits of D for all classes of coastal categorization and their characteristics are summarized in Table 4.10.

The top 85th percentile and lower 15th percentile of the D values on a normal (Gaussian) plot corresponds to Class I and Class V sites respectively (Ergin et al. 2004). With further contributions of additional public inquiries and evaluation of more coastal sites, classification borders may be subject to change, which reflects the dynamic and enhancing state of the methodology.

Examples of Class I sites ($D \geq 0.85$) included Phasalis Large Bay (Turkey), Dingle Cliffs (Malta), Poppit (UK) and Zlatni Rat (Croatia). These sites rated highly with their outstanding features represented by physical and human parameters. From the perception studies, the top five rated parameters obtained were water colour and clarity absence of sewage and litter, quality of built environment, absence of noise and as for the natural parameters beach type and natural vegetation cover.

Examples of Class II sites ($0.85 > D \geq 0.65$) included Fungus Rock (Malta), Nash Point (UK), Karaburun Akyar Mersin (Turkey) and Stiniva (Croatia). Fungus Rock was scored highly on spectacular cliff scenery yet beaches and natural vegetation cover rating lowered the overall assessment. The Nash Point site's lower rating was mainly due to water colour and clarity, lower landscape futures and a pebble beach (rather than sand). The Karaburun Akyar, Mersin site's lower rating was influenced by mainly intensive housing and tourism development out of harmony with nature. Stiniva, located in superb area with cliffs, a pebble beach, excellent water was rated high but its immense litter problem has lowered its rating.

Examples of Class III sites $(0.65 > D \geq 0.40)$ included Tenby South (UK), Ghajn Tuffieha (Malta), Göksu Hurma (Turkey) and Komiza 2 (Croatia). The Tenby South rating was lowered by poor water colour and clarity and a macro tidal range in spite of the positive features of the site including sensitive urban development. At Ghajn Tuffieha a very narrow sand beach with considerable amount of litter were the main negative aspects preventing a higher classification in spite of an absence of noise and intensive urban development and good natural vegetation. Finally, Göksu Hurma site, located at the mouth of a river delta, was rated as Class III due to considerable amount of litter and difficulty of access. For Komiza 2, physical factors scored well but absence of buffer zone has lowered its class.

Examples of Class IV sites $(0.40 > D \geq 0.00)$ included the largest number of sites. Examples of Class IV were Konyaaltı middle (Turkey), Kercem Cliffs (Malta), Saundersfoot (UK) and Supetar (Croatia). In all these sites intensive urban development with its associated problems of utilities, litter noise disturbance and a degeneration of natural features and vegetation cover were the dominating parameters.

Examples of Class V sites $(D < 0.00)$ included Kızkalesi, Mersin (Turkey), Amroth (UK), St Georges Bay (Malta) and Bacvice (Crotia). These sites were assessed with their intensive and extensive unattractive urbanization and tourism with ugly coastal structures, high amounts of litter and noise and degraded natural environments.

CSES enabled calculation of an Evaluation Index (D) which categorized the scenery of coastal sites evaluated and statistically best described attribute values in terms of weighted areas. A generalization seems possible for the physical and human parameters assessed for the coastal sites. Figure 4.9 shows the grades given to coastal scenic parameters by the experts (attributes 1 to 5) versus the D-value classification (I to V) of the coastal sites (Ergin et al. 2011). The dominant grades for the Class I and II coastal sites are definitely due to human parameters, whereas, for Class V coastal sites, physical parameters slightly prevail. For Classes I and II the expert grades for the human parameters are larger than 4.5 and the physical parameter grades are less than 3, whereas, for Class V coastal sites, the expert grades for the human and physical parameters are the lowest, and both are around 2.5. In all classes except Class V grading of the human parameters were reflected effectively in the final classification of the coastal sited as given by (D) values.

In conclusion, it can be stated that higher D-valued coastal sites have significantly greater grades for their human parameters than for their physical parameters, as checked by the experts and by public inquiries, compared to low expert grades for the physical and human parameters of Class V. In the context of coastal management policies, high human parameters at low attribute values reflect an intensive urban development, litter etc. Most sites will have physical parameters which would tend to constrain an increase in scenic value from the physical parameters. In this case emphasis should be given by coastal managers to find out ways of increasing human parameter scores.

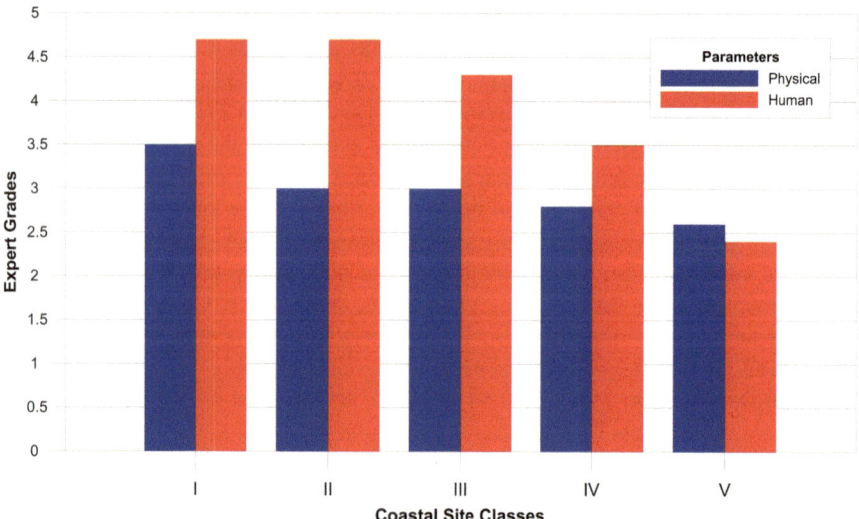

Fig. 4.9 Comparison of Coastal Scenic Parameter Grades in Coastal Classification (Ergin et al. 2011)

4.4 Methodology Validation: Simulation Work on Çıralı Karaburun

You know my methods, Watson. (A Conan Doyle, the Crooked Mat)

According to the "International Union for Conservation of Nature (IUCN) Categorizing System" Çıralı Karaburun is classified as a protected site corresponding to "Protected Landscape/Seascape" status. This site was assessed by CSES as a Class I (D = 1.26), where the expert grades were obtained as 3.44 and 4.62 out of 5 for physical and human parameters respectively. In the context of coastal protection policies Çıralı Karaburun was selected as the most suitable site among the Turkish coastal sites to carry out a simulation work for possible future development scenarios. Çıralı Karaburun has a high scenic value because of its good physical and especially human parameters yet is under threat as a result of the forcing function of human usage. Although at present, Çıralı Karaburun is a protected area, this status may not persist if regulations and rules are not obeyed and protection is weakened with regard to human usage. With its dominating high rated human parameters this site was suitable to make simulations using the CSES methodology. Long term scenarios were hypothetically generated using the CSES technique. With regard to alternative coastal zone management plans this simulation work was suitable for evaluating future potential changes in view of preservation and conservation and

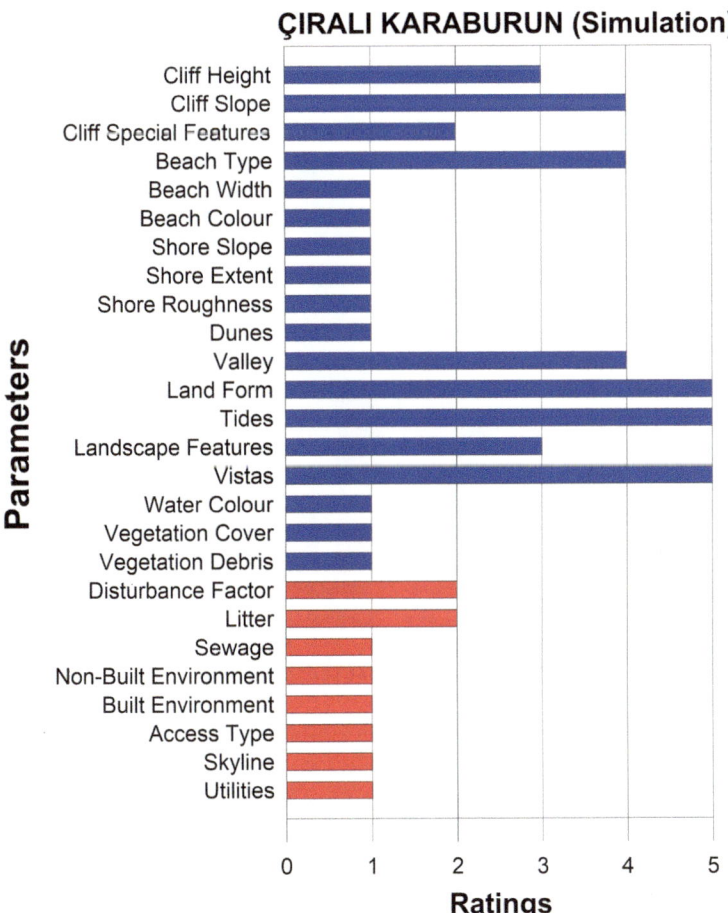

Fig. 4.10 Assessment Histogram simulated for Çıralı Karaburun

sustainable development of the coastal areas especially with regard to anthropogenic influences (e.g. built environment, litter, sewage, etc.) (Gezer 2004).

The simulation work very effectively reflected the adverse impacts of long term
future changes on the human parameters (e.g. intensive urban, tourism, industrial
development) consequently lowering the attribute values of physical parameters
selected as the beach width, beach color, water color, vegetation cover and vegetation debris (BCR 2004; Ergin et al. 2005, Ergin 2009).

Results of simulations for Çıralı Karaburun are presented in figures and assessment histogram (Fig. 4.10), weighted averages histograms for physical and human
parameters (Fig. 4.11) and membership degree graph (Fig. 4.12). Figures 4.10 and
4.11 reflects the adverse impact of low attribute values of human parameters. With
the membership degree graph having a left-hand skew curve (Fig. 4.12), the original

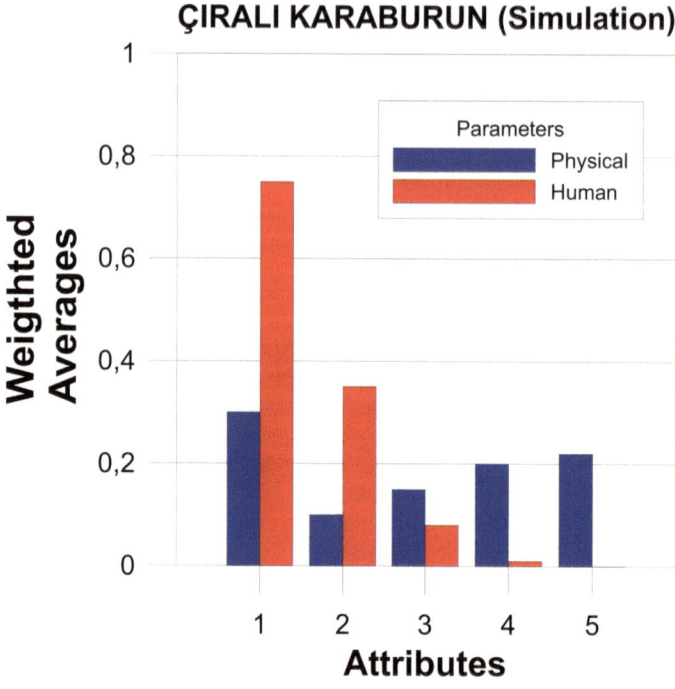

Fig. 4.11 Weighted Averages versus Attributes simulated for Çıralı Karaburun

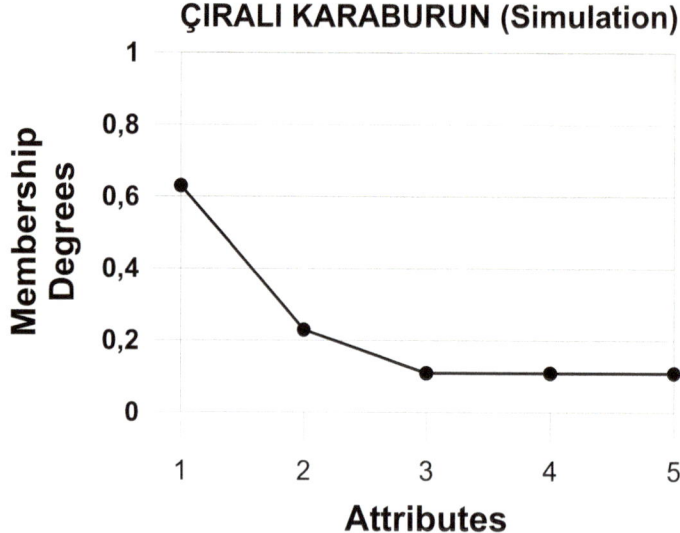

Fig. 4.12 Membership Degree *vs* Attributes simulated for Çıralı Karaburun

D-value of Çıralı Karaburun has decreased from D = 1.26 to D = −0.48 and has changed the state of the site from Class I to Class V.

The coastal scenic assessment work was utilized as guidelines for strategies for the conservation and management and land use plans of Çıralı Karaburun. At present, the site retains its Class I position through appropriate implementation of land use and management, successfully enabled by an environmental-friendly coastal usage. The land use plan has met with endorsement from the local community; local government institutions and the relevant ministries ensured the better managed cultural and natural resources. Diversified economic activities (tourism, organic agriculture, non-timber forest products) for environmentally and social sound development were successfully promoted. More importantly ecotourism has been taken up enthusiastically by the community, and the town has become a famous nature-friendly tourism destination. A sustainable tourism business (high occupancy rate, decent prices, and longer season) is booming. Çıralı Karaburun has a strong competition capability with neighboring mass tourism areas, by the help of the high quality tourism services provided. Organic agriculture is well adapted and 'The Çıralı Brand' is becoming a household name for high-quality organic products.

Improved protection of the marine turtle habitat, demonstrated by an increase in the number of nests, implies a positive effect on the marine and coastal biodiversity. Another protection component consisted of educational activities to heighten local residents' awareness of natural values.

Çıralı Karaburun stands as a very good example of utilizing the results of the CSES innovative methodology as a useful guideline for assessing, designing and managing coastal areas when future plans must be made regarding intensive urban and industrial development.

4.5 Conclusions

Joy in looking and comprehending is Nature's most beautiful gift. (Albert Einstein)

Although scenery is an invaluable asset from any environmental viewpoint, coastal scenery is a lesser-considered aspect of coastal management. In this regard, an innovative scenic evaluation CSES technique was developed using fuzzy logic methodology to evaluate the adverse effects of changes to a coastal environment.

CSES comprising 18 physical and 8 human scenic parameters initially assessed coastal sites from U.K., Turkey, Malta, and Croatia on a five scale attribute rating system. Membership degree versus attribute curves, were used to develop for the most appropriate evaluation index (D) criteria. The skew of the membership curves reflected the scenic value of the assessed sites.

An evaluation index curve D was obtained for 90 coastal sites (now evaluated for 952 world sites) which enabled five classes of scenery to be differentiated. It should be emphasized that not only the weights of the parameters but also the attributes scored for each parameter (especially for the human factors) will significantly change

D values. The effectiveness of the human parameters on D values is a good example of the threat of a forcing function for planners and managers in assessing human input.

Coastal scenery, as a part of coastal landscape inventory available for planners and managers may be utilized as a functional tool for assessment for coastal preservation, protection and sustainable development. Yet it must be clearly understood that, site assessment is time dependent so the D value of a site is not an intrinsic quantity. It is well known that, although many physical features of a coast change very slowly, the human parameters, such as noise disturbance, litter, the built environment, and the type of access, may change considerably in a short time. Therefore, coastal assessment surveys are recommended to be carried out periodically yielding the changes in the classification of the coastal sites, which can be used to guide coastal planners, managers, and designers. Sensitive designs in all respects and buffer zone applications will definitely change the present status of a coast, as specified by the D values discussed here. Coastal zones are very important especially with respect to global warming, resulting in adverse effects on coastal areas from events such as sea level rise, erosion, inundation, and flooding. In that respect to make sound decisions about coastal zones even more important.

In conclusion, CSES may be used as a tool for making decisions on the preservation, conservation, and sustainable development of coastal areas. Primary concern for relevant planners and managers should be on highly appreciated parameters, such as the existence of buffer zones, and the absence of litter and sewage, as well as the building environment. Coastal sites determined by CSES to be Class I and Class V can be used as models to encourage future coastal planners to respect a site's natural values and to avoid intensive urban and industrial development.

CSES Open-Source Computational Tool

Coastal scenic evaluation system (CSES) is re-implemented in MATLAB environment in Ergin et al. (2018). The developed code 'CSES 2018 V1' is presented as an open-source computational tool for the coastal scenic evaluation system with User's Manual and with an example case study on the following website: http://cses.ce.metu.edu.tr/

Addendum

For different coast sites the results of final assessment matrices (R) obtained depend on membership grading via membership matrices and weights of parameters as well as graded attributes. The entry with maximum entry value of the final assessment matrix R, as membership degree can be accepted as an evaluation state of the coastal site. Yet, considering all sites together for further comparison and for normal (Gaussian) analysis a D-value is defined. D-Values to categorize the scenery of coastal sites are evaluated on statistically described attribute values in terms of weighted areas. So, D value does not change linearly with respect to change of attributes directly when weights of parameters and membership grading remain constant for some situations. Since, D values are evaluated on areas formed by the entries of the final assessment matrices where ticked attributes with 1 and 5 are given weights as –2 and 2, respectively. The ticked attributes 3 and 4 have assumed weights as –1 and 1; and for ticked attribute 3, weight is 0. The idea here is to enhance the results on both ends firstly for preservation and protection of the coastal sites.

D – Values defined in this study depend on weights, and the membership matrices, M_i of the parameters obtained after expert decisions during the BCR and are listed in Appendix 1. Site assessments are time dependent, so D- value is not an intrinsic quantity. They are open to changes in future with further researches and coastal assessment surveys.

Appendices

Appendix 1: Membership-Grade Matrices (Mi) of the 26 Scenic Parameters

Scenic Parameters

M	Physical parameters	
1	Cliff	Height (*H*)
2		Slope
3		Special features
4	Beach face	Type
5		Width (*W*)
6		Colour
7	Rocky shore	Slope
8		Extent
9		Roughness
10	Dunes	
11	Valley	
12	Skyline landforms	
13	Tides	
14	Coastal landscape features	
15	Vistas	
16	Water colour & clarity	
17	Vegetation cover	

(continued)

18	Vegetation debris
	Human parameters
19	Disturbance factor (noise)
20	Litter
21	Sewage (discharge evidence)
22	Non-built environment
23	Built environment
24	Access type
25	Skyline
26	Utilities

Appendix 2: Membership-Grade Matrices

$$M_1 = \begin{bmatrix} 1.0 & 0.0 & 0.0 & 0.0 & 0.0 \\ 0.0 & 1.0 & 0.3 & 0.0 & 0.0 \\ 0.0 & 0.3 & 1.0 & 0.3 & 0.0 \\ 0.0 & 0.0 & 0.5 & 1.0 & 0.5 \\ 0.0 & 0.0 & 0.0 & 0.5 & 1.0 \end{bmatrix} \quad M_2 = \begin{bmatrix} 1.0 & 0.0 & 0.0 & 0.0 & 0.0 \\ 0.0 & 1.0 & 0.5 & 0.0 & 0.0 \\ 0.0 & 0.5 & 1.0 & 0.5 & 0.0 \\ 0.0 & 0.0 & 0.5 & 1.0 & 0.5 \\ 0.0 & 0.0 & 0.0 & 0.5 & 1.0 \end{bmatrix}$$

$$M_3 = \begin{bmatrix} 1.0 & 0.0 & 0.0 & 0.0 & 0.0 \\ 0.0 & 1.0 & 0.3 & 0.0 & 0.0 \\ 0.0 & 0.0 & 1.0 & 0.3 & 0.0 \\ 0.0 & 0.0 & 0.0 & 1.0 & 0.3 \\ 0.0 & 0.0 & 0.0 & 0.0 & 1.0 \end{bmatrix} \quad M_4 = \begin{bmatrix} 1.0 & 0.0 & 0.0 & 0.0 & 0.0 \\ 0.0 & 1.0 & 0.0 & 0.0 & 0.0 \\ 0.0 & 0.0 & 1.0 & 0.0 & 0.0 \\ 0.0 & 0.0 & 0.0 & 1.0 & 0.0 \\ 0.0 & 0.0 & 0.0 & 0.0 & 1.0 \end{bmatrix}$$

(continued)

$$M_5 = \begin{bmatrix} 1.0 & 0.0 & 0.0 & 0.0 & 0.0 \\ 0.0 & 1.0 & 0.0 & 0.0 & 0.0 \\ 0.0 & 0.2 & 1.0 & 0.2 & 0.0 \\ 0.0 & 0.0 & 0.2 & 1.0 & 0.6 \\ 0.0 & 0.0 & 0.0 & 0.6 & 1.0 \end{bmatrix}$$

$$M_6 = \begin{bmatrix} 1.0 & 0.0 & 0.0 & 0.0 & 0.0 \\ 0.0 & 1.0 & 0.0 & 0.0 & 0.0 \\ 0.0 & 0.0 & 1.0 & 0.6 & 0.0 \\ 0.0 & 0.0 & 0.6 & 1.0 & 0.0 \\ 0.0 & 0.0 & 0.0 & 0.0 & 1.0 \end{bmatrix}$$

$$M_7 = \begin{bmatrix} 1.0 & 0.0 & 0.0 & 0.0 & 0.0 \\ 0.0 & 1.0 & 0.5 & 0.0 & 0.0 \\ 0.0 & 0.5 & 1.0 & 0.5 & 0.0 \\ 0.0 & 0.0 & 0.5 & 1.0 & 0.5 \\ 0.0 & 0.0 & 0.0 & 0.2 & 1.0 \end{bmatrix}$$

$$M_8 = \begin{bmatrix} 1.0 & 0.0 & 0.0 & 0.0 & 0.0 \\ 0.0 & 1.0 & 0.2 & 0.0 & 0.0 \\ 0.0 & 0.2 & 1.0 & 0.5 & 0.0 \\ 0.0 & 0.0 & 0.5 & 1.0 & 0.4 \\ 0.0 & 0.0 & 0.0 & 0.4 & 1.0 \end{bmatrix}$$

$$M_9 = \begin{bmatrix} 1.0 & 0.0 & 0.0 & 0.0 & 0.0 \\ 0.0 & 1.0 & 0.1 & 0.0 & 0.0 \\ 0.0 & 0.1 & 1.0 & 0.6 & 0.0 \\ 0.0 & 0.0 & 0.6 & 1.0 & 0.5 \\ 0.0 & 0.0 & 0.0 & 0.5 & 1.0 \end{bmatrix}$$

$$M_{10} = \begin{bmatrix} 1.0 & 0.0 & 0.0 & 0.0 & 0.0 \\ 0.0 & 1.0 & 0.0 & 0.0 & 0.0 \\ 0.0 & 0.0 & 1.0 & 0.0 & 0.0 \\ 0.0 & 0.0 & 0.0 & 1.0 & 0.0 \\ 0.0 & 0.0 & 0.0 & 0.0 & 1.0 \end{bmatrix}$$

$$M_{11} = \begin{bmatrix} 1.0 & 0.0 & 0.0 & 0.0 & 0.0 \\ 0.0 & 1.0 & 0.0 & 0.0 & 0.0 \\ 0.0 & 0.0 & 1.0 & 0.0 & 0.0 \\ 0.0 & 0.0 & 0.0 & 1.0 & 0.1 \\ 0.0 & 0.0 & 0.0 & 0.1 & 1.0 \end{bmatrix}$$

$$M_{12} = \begin{bmatrix} 1.0 & 0.2 & 0.0 & 0.0 & 0.0 \\ 0.0 & 1.0 & 0.3 & 0.0 & 0.0 \\ 0.0 & 0.6 & 1.0 & 0.6 & 0.0 \\ 0.0 & 0.0 & 0.6 & 1.0 & 0.2 \\ 0.0 & 0.0 & 0.0 & 0.2 & 1.0 \end{bmatrix}$$

$$M_{13} = \begin{bmatrix} 1.0 & 0.0 & 0.0 & 0.0 & 0.0 \\ 0.0 & 0.0 & 0.0 & 0.0 & 0.0 \\ 0.0 & 0.0 & 1.0 & 0.0 & 0.0 \\ 0.0 & 0.0 & 0.0 & 0.0 & 0.0 \\ 0.0 & 0.0 & 0.0 & 0.0 & 1.0 \end{bmatrix}$$

$$M_{14} = \begin{bmatrix} 1.0 & 0.2 & 0.0 & 0.0 & 0.0 \\ 0.0 & 1.0 & 0.2 & 0.0 & 0.0 \\ 0.0 & 0.0 & 1.0 & 0.2 & 0.0 \\ 0.0 & 0.0 & 0.0 & 1.0 & 0.2 \\ 0.0 & 0.0 & 0.0 & 0.0 & 1.0 \end{bmatrix}$$

$$M_{15} = \begin{bmatrix} 1.0 & 0.0 & 0.0 & 0.0 & 0.0 \\ 0.0 & 1.0 & 0.0 & 0.0 & 0.0 \\ 0.0 & 0.0 & 0.0 & 0.0 & 0.0 \\ 0.0 & 0.0 & 0.0 & 1.0 & 0.3 \\ 0.0 & 0.0 & 0.0 & 0.3 & 1.0 \end{bmatrix}$$

$$M_{16} = \begin{bmatrix} 1.0 & 0.2 & 0.0 & 0.0 & 0.0 \\ 0.2 & 1.0 & 0.2 & 0.0 & 0.0 \\ 0.0 & 0.5 & 1.0 & 0.5 & 0.0 \\ 0.0 & 0.0 & 0.5 & 1.0 & 0.2 \\ 0.0 & 0.0 & 0.0 & 0.2 & 1.0 \end{bmatrix}$$

(continued)

$$M_{17} = \begin{bmatrix} 1.0 & 0.2 & 0.0 & 0.0 & 0.0 \\ 0.2 & 1.0 & 0.2 & 0.0 & 0.0 \\ 0.0 & 0.2 & 1.0 & 0.2 & 0.0 \\ 0.0 & 0.0 & 0.2 & 1.0 & 0.2 \\ 0.0 & 0.0 & 0.0 & 0.2 & 1.0 \end{bmatrix} \quad M_{18} = \begin{bmatrix} 1.0 & 0.2 & 0.0 & 0.0 & 0.0 \\ 0.2 & 1.0 & 0.0 & 0.0 & 0.0 \\ 0.0 & 0.0 & 1.0 & 0.2 & 0.0 \\ 0.0 & 0.0 & 0.2 & 1.0 & 0.0 \\ 0.0 & 0.0 & 0.0 & 0.2 & 1.0 \end{bmatrix}$$

$$M_{19} = \begin{bmatrix} 1.0 & 0.0 & 0.0 & 0.0 & 0.0 \\ 0.2 & 1.0 & 0.0 & 0.2 & 0.0 \\ 0.0 & 0.0 & 0.0 & 0.0 & 0.0 \\ 0.0 & 0.2 & 0.0 & 1.0 & 0.2 \\ 0.0 & 0.0 & 0.0 & 0.2 & 1.0 \end{bmatrix} \quad M_{20} = \begin{bmatrix} 1.0 & 0.2 & 0.0 & 0.0 & 0.0 \\ 0.2 & 1.0 & 0.2 & 0.0 & 0.0 \\ 0.0 & 0.2 & 1.0 & 0.2 & 0.0 \\ 0.0 & 0.0 & 0.2 & 1.0 & 0.2 \\ 0.0 & 0.0 & 0.0 & 0.2 & 1.0 \end{bmatrix}$$

$$M_{21} = \begin{bmatrix} 1.0 & 0.0 & 0.2 & 0.0 & 0.0 \\ 0.0 & 0.0 & 0.0 & 0.0 & 0.0 \\ 0.3 & 0.0 & 1.0 & 0.0 & 0.1 \\ 0.0 & 0.0 & 0.0 & 0.0 & 0.0 \\ 0.0 & 0.0 & 0.2 & 0.0 & 1.0 \end{bmatrix} \quad M_{22} = \begin{bmatrix} 1.0 & 0.0 & 0.2 & 0.0 & 0.0 \\ 0.0 & 0.0 & 0.0 & 0.0 & 0.0 \\ 0.2 & 0.0 & 1.0 & 0.0 & 0.2 \\ 0.0 & 0.0 & 0.0 & 0.0 & 0.0 \\ 0.0 & 0.0 & 0.2 & 0.0 & 1.0 \end{bmatrix}$$

$$M_{23} = \begin{bmatrix} 1.0 & 0.0 & 0.0 & 0.0 & 0.0 \\ 0.0 & 1.0 & 0.2 & 0.0 & 0.0 \\ 0.0 & 0.2 & 1.0 & 0.2 & 0.0 \\ 0.0 & 0.0 & 0.3 & 1.0 & 0.0 \\ 0.0 & 0.0 & 0.0 & 0.0 & 1.0 \end{bmatrix} \quad M_{24} = \begin{bmatrix} 1.0 & 0.2 & 0.0 & 0.0 & 0.0 \\ 0.2 & 1.0 & 0.0 & 0.2 & 0.0 \\ 0.0 & 0.0 & 0.0 & 0.0 & 0.0 \\ 0.0 & 0.2 & 0.0 & 1.0 & 0.2 \\ 0.0 & 0.0 & 0.0 & 0.2 & 1.0 \end{bmatrix}$$

$$M_{25} = \begin{bmatrix} 1.0 & 0.4 & 0.0 & 0.0 & 0.0 \\ 0.4 & 1.0 & 0.2 & 0.0 & 0.0 \\ 0.0 & 0.4 & 1.0 & 0.2 & 0.0 \\ 0.0 & 0.0 & 0.4 & 1.0 & 0.0 \\ 0.0 & 0.0 & 0.0 & 0.0 & 1.0 \end{bmatrix} \quad M_{26} = \begin{bmatrix} 1.0 & 0.0 & 0.0 & 0.0 & 0.0 \\ 0.2 & 1.0 & 0.0 & 0.0 & 0.0 \\ 0.0 & 0.2 & 1.0 & 0.0 & 0.0 \\ 0.0 & 0.0 & 0.2 & 1.0 & 0.0 \\ 0.0 & 0.0 & 0.0 & 0.2 & 1.0 \end{bmatrix}$$

References

Appleton J (1975) Landscape evaluation: the theoretical vacuum. Trans Inst Br Geology 66:120–124

BCR (British Council Report) (2003) Coastal scenic assessment at selected areas: Turkey, UK, Malta. British council project final report, Ankara by Ergin, A., Williams, A.T., Micallef, A

BCR (British Council Report) (2004) A new methodology for evaluating coastal scenery: fuzzy logic systems –a pilot study for Cirali. British Council Project Final Report, Ankara, by Ergin A, Williams AT, Micallef A

Briggs DJ, France J (1980) Landscape evaluation: a comparative study. J Environ Manag 10:263–275

Buyoff GJ, Arndt LK (1981) Interval scaling of landscape preference by direct and indirect measurement methods. Landsc Plan 8:257–267

Carlson AA (1977) On the possibility of quantifying scenic beauty. Landscape. Planning 4:131–172

CCW (Countryside Council for Wales) (1996) The Welsh Landscape: our heritance and its future protection and enhancement

CCW (Countryside Council for Wales) (2001) The LANDMAP Information System

Clamp PE (1976) Evaluating English landscapes. Environ Plan A8:79–92

Countryside Commission (1987) Landscape assessment-a countryside commission approach. 18, 143pp

Countryside Commission (1993) Landscape Assessment Guidance. CCP423, CC. Cheltenham, Gloc. UK

Dakin S (2003) There is more to landscape than meets the eye: towards inclusive landscape assessment in resource and environmental management. Can Geogr 47(2):185–200

Daniel TC (1990) Measuring the quality of the natural environment: a psychophysical approach. Am Psychol 45(5):633–637

Elettheriadis N, Tsalikidis I, Manos B (1990) Coastal landscape preference evaluation: a comparison among tourists in Greece. Environ Manag 14(4):475–487

Ergin A, Karaesmen E, Uçar B (2011) A quantitative study for evaluation of coastal scenery. J Coast Res 27(6):1065–1075

Ergin A (2009) Case study; a holistic approach to beach management at Çıralı, Turkey: a model of conservation, integrated management and sustainable development. In: Williams AT, Micallef A (eds) Beach management: principles and practice. Earthscan, London, pp 355–358

Ergin A, Williams AT, Micallef A (2006) Coastal scenery: appreciation and evaluation. J Coast Res 22:958–964

Ergin A, Karaesmen E, Gezer E, Uçar B, Karakaya ST (2005) Çıralı coastal scenic assessment using fuzzy logic mathematics. KAY Symposium 2005 (in Turkish)

Ergin, A., Karaesmen, E., Guler, I., Guler, H. G. (2018) "Development of An Open-Source Computational Tool for Coastal Scenic Assessment Based on Fuzzy Logic," 9th Coastal Engineering Symposium Proceedings, Turkish Chamber of Civil Engineers, Adana, Turkey.

Ergin A, Karaesmen E, Williams AT, Micallef A (2004) A new methodology for evaluating coastal scenery: fuzzy logic systems. Area 36(4):367–386

Ergin A, Williams AT, Micallef A, Karakaya ST (2002) An innovative approach to coastal scenic evaluation Beach Management in the Mediterranean and Black Sea. MEDCOAST METU, Ankara, pp 215–226

Fines KD (1968) Landscape evaluation. A research project in East Sussex. Reg Stud 2:41–55

Gezer E (2004) Coastal scenic evaluation: a pilot study for Çıralı. MSc thesis, Middle East Technical University, Ankara, Turkey. 110 p

Kaplan R, Kaplan S (1989) The visual environment: public participation in design and planning. J Soc Issue 45(1):59–86

Karakaya ST (2004) Coastal scenic assessment using fuzzy logic approach. Middle East Technical University, Ankara, Turkey, MSc Thesis, 164 p

Leopold LB (1969) Quantitative comparisons of some aesthetic factors among rivers. US Geol Survey Circ 620. 16 pp

Linton DL (1968) The assessment of scenery as a natural resource. Scott Geogr Mag 84:219–238

Linton DL (1982) Visual assessments of natural landscapes. West Geogr Ser 20:97–116

Lowenthal D (1967) Environmental Perception and Behaviour [Edited versions of papers presented in a symposium at the 61st annual meeting of the Association of American Geographers, Columbus, Ohio, April 20, 1965]. University of Chicago Press, Chicago. 102 p

Lowenthal D (1978) Finding valued landscapes. Institute of Environmental Sciences/University of Toronto, Toronto

Penning-Rowsell EC (1982) A public preference evaluation of landscape quality. Reg Stud 16:97–112

Penning-Rowsell EC (1989) Landscape evaluation in practice – a survey of local authorities. Landsc Res 14(2):35–37

Robinson DG (1976) Landscape Evolution: the Landscape Evaluation Research Project. 1970–75, Manchester University, UK

Uçar B (2004) Coastal scenic evaluation by application of fuzzy logic mathematics. MSc thesis. Middle East Technical University, Ankara, Turkey. 115 p

Van der Meulen F (1997) History and Culture as expressed in the landscape seen from an Ecological Viewpoint. Science, Religion and the Environment Symposium II: The Black Sea in Crisis, pers comm

Williams AT, Khatabi A (2015) Beach scenery, Nador, Morocco. J Coast Conserv 19(5):743–755

Zadeh LA (1965) Fuzzy sets. Inf Control 8:338–335

Chapter 5
Coastal Scenery Assessment: Definitions and Typology

Enzo Pranzini, Allan T. Williams, and Nelson Rangel-Buitrago

Abstract Although scenery is an invaluable asset from any environmental viewpoint, coastal scenery is a lesser-considered aspect of coastal management. Therefore the Coastal Scenic Evaluation System (CSES) technique was developed using fuzzy logic methodology to evaluate the adverse effects of changes to a coastal environment. The CSES can be used not only for landscape preservation and protection, but also as scientific tool for envisaged coastal management and future development based upon plans by an evidence-based approach. This chapter presents a detailed field guide that includes the steps to follow in making a coastal scenery assessment, as well as, key definitions and examples of all parameters required to use the CSES. The Photo-atlas representing all parameters helps in attributing them to the correct grade.

5.1 How to Use the Coastal Scenic Evaluation System (CSES)

The coastal scenery assessment is carried out by means of a five scale attribute rating system that evaluates 18 physical and 8 human parameters defined into the "Coastal Scenic Evaluation System" (Table 5.1; see Chap. 4 for a detailed discussion of this table).

E. Pranzini (✉)
Department of Earth Sciences, University of Florence, Florence, Italy
e-mail: enzo.pranzini@unifi.it

A. T. Williams
Faculty of Architecture, Computing and Engineering, University of Wales, Trinity Saint David, Swansea, Wales, UK

CICA NOVA, Nova Universidad de Lisboa, Lisboa, Portugal

N. Rangel-Buitrago
Departamentos de Física y Biologia, Facultad de Ciencias Básicas, Universidad del Atlántico, Barranquilla, Atlántico, Colombia
e-mail: nelsonrangel@mail.uniatlantico.edu.co

© Springer International Publishing AG, part of Springer Nature 2019   107
N. Rangel-Buitrago (ed.), *Coastal Scenery*, Coastal Research Library 26,
https://doi.org/10.1007/978-3-319-78878-4_5

Table 5.1 Coastal Scenic Evaluation System (Ergin et al. 2004)

N	Physical Parameters		Rating				
			1	2	3	4	5
1	Cliff	Height (H)	Absent (< 5 m)	5 m ≤ H < 30 m	30 m ≤ H < 60 m	60 m ≤ H < 90 m	H ≥ 90 m
2		Slope	< 45°	45° – 60°	60° – 75°	75° – 85°	Circa vertical
3		Special features[a]	Absent	1 special feature	2 special features	3 special features	Many > 3 special features.
4	Beach face	Type	Absent	Mud	Cobble/boulder	Pebble/gravel	Sand
5		Width (W)	Absent	W < 5 m or W > 100 m	5 m ≤ W < 25 m	25 m ≤ W < 50 m	50 m ≤ W ≤ 100 m
6		Colour	Absent	Dark	Dark tan	Light tan/bleached	White/gold
7	Rocky shore	Slope	Absent	< 5°	5° – 10°	10° – 20°	> 20°
8		Extent	Absent	< 5 m	5 m – 10 m	10 m – 20 m	> 20 m
9		Roughness	Absent	Distinctly jagged	Deeply pitted and/or irregular	Shallow pitted	Smooth
10	Dunes		Absent	Remnants	Fore-dune	Secondary ridge	Several
11	Valley		Absent	Dry	Stream (< 1 m)	Stream (1 m – 4 m)	> 4 m
12	Skyline landforms		Not visible	Flat	Undulating	Highly undulating	Mountainous
13	Tides		Macro (> 4 m)		Meso (2 m – 4 m)		Micro (< 2 m)
14	Coastal landscape features[b]		None	1 feature	2 features	3 features	>3 features
15	Vistas		Open on one side	Open on two sides		Open on three sides	Open on four sides
16	Water colour & clarity		Muddy Brown/grey	Milky blue/green; opaque	Green/grey blue	Clear blue/dark blue	Very clear turquoise
17	Vegetation cover		Bare (< 10% vegetation only)	Scrub/Garigue/grass (marram/ferns, bramble/meadow, etc)	Wetland/meadow	Coppices, maquis (mature trees bushes)	Variety of mature trees/mature natural cover
18	Vegetation debris		Continuous > 50 cm high	Full strand line	Single accumulation	Few scattered items	None

Human parameters

19	Disturbance factor (noise)	Intolerable	Tolerable		Little	None
20	Litter	Continuous accumulations	Full strand line	Single accumulation	Few scattered items	Virtually absent
21	Sewage (discharge evidence)	Sewage evidence		Some sewage evidence		No evidence of sewage
22	Non-built environment[c]	None		Hedgerow/terracing/monoculture		Field mixed cultivation ± trees/natural
23	Built environment[d]	Heavy industry	Heavy tourism and/or urban	Light tourism and/or urban and/or sensitive industry	Sensitive tourism and/or urban	Historic and/or none
24	Access type	No buffer zone/heavy traffic	Buffer zone/light traffic		Parking lot visible from coastal area	Parking lot not visible from coastal area
25	Skyline	Very unattractive	Unattractive	Sensitively designed	Very sensitively designed	Natural/historic features
26	Utilities[e]	>3 utilities	3 utilities	2 utilities	1 utility	None

[a]Cliff special features: Indentation, banding, folding, scree, irregular profile, etc.

[b]Coastal landscape features: Peninsulas, rock ridges, irregular headlands, arches, windows, caves, waterfalls, deltas, lagoons, islands, stacks, estuaries, reefs, fauna, embayment, tombola, etc.

[c]Non-built environment: When there is no agricultural activity.
If the natural vegetation cover parameter (17) has scored a 5, then tick the 5 box here.
If the natural vegetation cover parameter (17) has scored 2, 3, 4 then tick the 3 box here.

[d]Built environment: Caravans will come under tourism, grading 2: Large intensive caravan site, grading 3: Light, but still intensive caravan sites, grading 4: Sensitively designed caravan sites

[e]Utilities: Power lines, pipelines, street lamps, groins, seawalls, revetments, etc.

To fill in the scenic evaluation checklist an assessor(s) visits a specific coastal location between 10 am and 3 pm under normal weather conditions for the site and over a 100 m range along the site ticks off relevant parameter boxes given in Table 5.1. This allows determination of the site state according to its current attributes.

As an example, the Valley parameter has five attributes: absent, dry valley, stream (width <1 m), stream (width 1 m–4 m) and river (width >4 m). A coastal site with a dry valley would be scored as 2; a site with a river would be scored as 5. Although some parameters can be easily quantified (such as beach width, number of utilities etc.), several used in this classification are subject to the perception over the coastal site, e.g. water colour, built environment, that is why it is semi –quantitative. The background to Table 5.1 (see Chap. 4) is the key to scenic assessment and given in detail below are definitions and figures corresponding to the parameters given in Table 5.1.

5.2   Parameters, Definition and Scoring

(Numbers identifies parameters both on Table 5.1 and the Photo-atlas in the Appendix of this chapter).

1–3 Cliff
Is a steeply (>45°) sloping surface where elevated land (>5 m) meets the shoreline. These two items are mandatory (>5 m in height and a slope >45°) and the cliff may be composed of rock and/or unconsolidated material or boulder clay.

1 Height   is measured from the base or from the water line to the slope break (where slope is <45°).

2 Slope   The slope angle depends on a variety of factors including jointing, bedding and hardness of the cliff materials, as well as the erosional processes.

3 Special Features can include:

Banding. When cliffs are composed of various layers *e.g.* rock sedimentation units (i.e. alternate shale and limestone), or as well as, volcanic basalt layers and all undergo differential erosion. Various **colours** can differentiate the rock bands.

Faulting. Where Earth movements have displaced individual rock units a line can be seen (fault line) which has shifted layers on either side.

Folding. Where rocks have been under pressure and folded to accommodate the pressure. Folding can be gentle with a long wavelength to very severe with short tight wavelengths.

Gullying. In unconsolidated material rain action can erode material forming deep rills and then gullies.

Indentation. The shape/orientation of the cliff. It could be straight or curved, the greater the curvature the more highly indented is the cliff.

Scree/talus. Accumulation of rock material at the foot of, or mantling cliff slopes.

Tufa. Calcareous material deposits on a limestone cliff face due to water seepage.

Unconformity. A type of discontinuity. It represents a surface between successive strata indicating a missing geological time interval. It could be caused by an interruption in deposition and/or an erosional period followed by renewed deposition.

Dikes. Sheet of rock, usually hard, running across individual strata.

Sill. A tabular intrusion that has intruded between older layers of the cliff.

4–6 Beach Face

A deposit of non-cohesive material sited at the land/water interface and actively worked by waves, currents and sometimes wind.

4 Type Type relates to the non-cohesive material composing the beach e.g. **mud**, **sand**, **gravel**, **pebbles**. Mixed sediment beaches (sand + gravel; gravel + pebbles) are frequent; the dominant sediment must be considered.

5 Width (*W*) is the area between the water's edge and the back of a beach. The latter could be a wall, dune, building etc. The dry beach represents the area from high tide to rear of beach; intertidal zone equates to the low to high tide area. Measurement should be from the low **tide** position.

6 Color (e.g. **black**, **tan**, **white**, etc.) perceived by the human eye under full sun illumination.

7–9 Rocky Shore

Intertidal area in front of a cliff where physical processes (cliff retreat/wave action and/or haloclasty) predominates, usually producing a gentle sloping/horizontal rock platform.

7 Slope average slope measured from cliff face to the platform edge.

8 Extent Width is its extent from a cliff face to the platform edge.

9 Roughness is the degree of pitting on the surface, which varies from rough to smooth.

10 Dunes

An accumulation of wind-blown sand on the backshore, usually as small hummocks and swales, which tend to be stabilized by vegetation or control structures.

Fore-dunes represent the main dune adjacent to the beach frequently termed yellow dunes.

Secondary dune ridges are located behind the foredune and represent old fore-dunes that have been colonised by plants as the dune system extends seawards.

There may be several ridges and these are loosely called grey dunes.

11 Valley

A valley is a V shaped landscape feature formed by flowing water. If no water is present, it is termed a **dry valley**. If water is present the valley form can range from a small stream (**<1 m**) to a large river (**<4 m**). In fjord areas, glacial activity will have scoured the pre-existing river valley to a U shape.

12 Skyline Landforms

Looking landward the relief form can be estimated, i.e. **flat, undulating, highly undulating** or **mountainous.**

13 Tides

These are the result of the alternating rise and fall in sea level in oceans and seas produced by the gravitational attraction of the sun and more importantly, the moon. Tidal range refers to the height difference between the highest and lowest spring tide. The range for **Micro** tides is <2 m, **Meso** 2-4 m and **Macro** > 4 m.

14 Some Coastal Landscape Features

Distinctive forms related to erosional and/or depositional processes.

A **Peninsula/headland** is an area of land surrounded by water for the majority of its perimeter, but is connected to the mainland, i.e. it is land surrounded by water on three sides.

A **bay** is a recessed, coastal water body that connects directly to a larger main body of water, such as an ocean, lake, or another bay, i.e. it is surrounded by land on three sides and can be large or small.

A **cave** is a hollow in a cliff face that is caused usually by combinations of wave action, rock slippage, weathering, faulting, etc. Where a **cave** breaks through a cliff headland it forms an **arch**.

A **lagoon** is a shallow body of salt/fresh water separated from a larger body of water by barrier islands, shingle beaches, reefs or similar structures with an entrance to the sea.

A **sandbank** is a raised area of sand below the sea surface that can only be seen when the water level is at low **tide**.

A **sandbar** is permanently under water.

A **stack** is a steep, often vertical, sided column of rock rising from the sea. When part of a headland is eroded by hydraulic action and weathering, this can cause later collapse, forming free-standing stacks. Collapse of an **arch** can produce the same result.

A **tombolo** is a narrow depositional landform (usually sand or shingle) which connects an island to the mainland.

A **delta** is a low lying coastal plain of alluvial deposit protruding into the sea with a variety of shapes.

An **estuary** is a partially enclosed coastal body area of brackish water that has one or more rivers or streams flowing into it, i.e. it is the transition zone between salt/fresh water.

A **reef** is a hard structure located at or close to sea level. It can be a degraded **stack** or a narrow rock longshore barrier due to erosion and/or biogenic construction (corals, vermetids, etc.).

A **window** occurs if **cave**(s) carves through a headland above the water line resulting in a hole through the cliff.

15 Vistas

Are related to the line of sight to far off views, as a site could be enclosed on 1, 2, or 3 sides – the 4th side is always open to the sea. A far vista is where the foreground hill has another secondary background feature visible, e.g. a higher hill/mountain.

16 Water Colour & Clarity

The colour of the ocean is determined by interactions of incident light with substances or particles present in the water. The most significant constituents are free floating photosynthetic organisms (phyto-plankton) and inorganic particulates, the more there are of these the less the clarity.

17 Vegetation Cover

Is indicative of the flora of the coastal area vicinity close enough to visually affect a visitor's appreciation of the location. The emphasis is **natural**. Vegetation planted by humans, e.g. along promenades for a buffer zone is discounted.

18 Vegetation Debris

Can be brought to any location via land/sea sources. For example, seaweed (macroscopic, multicellular, marine algae) is a green, brown or dark red plant that grows almost exclusively in shallow waters at the ocean/sea edge or in the inter-tidal area. An excessive amount represents unattractive views to beach users. Similarly in the Mediterranean and other areas *Posidonia oceanica* leaves (a phanerogam), are common debris found on beaches. In tropical areas, sizable rivers bring large amounts of vegetation to the coast as a result of the abundant sun and rain in these locations.

19 Disturbance Factor (Noise)

Noise that may harm the activities developed at a coastal location, *e.g.* playing loud radio/CD music, jet skis, heavy traffic, airport noise, etc.

20 Litter

This is anthropogenic generated discards. Examples are beer cans, sweet wrappers, plastic bags, etc. along the beach, mostly found at the strand line or back of the beach. An accumulation represents these materials found in a mound.

21 Sewage (Discharge Evidence)

Relates to human/animal waste products, as well as its associated accessories, e.g. **sewage pipes** draining to beach, **condoms**, **tampon applicators**, **nappies**, etc.

22 Non-built Environment

The environment as seen minus its buildings. It relates to field patterns, agricultural, practices, etc.

23 Built Environment

Relates to any environment that includes anthropogenic structures, e.g. heavy/light industries, buildings, etc.

24 Access Type

A buffer zone is an area that divides two separate entities. For example, a tree lined promenade, or a natural grass area that separates a beach from a coastal road.

25 Skyline

The skyline silhouette of buildings that are or not in harmony with the environment, e.g. if building lines of the same height and blend into the environment. Discord exists if they stand out from the surroundings. If no buildings can be seen, then the area is in a natural state and attribute 5 is ticked in Table 5.1.

26 Utilities

These include items such as power lines, telephone lines/poles, lighting, communication masts, railways, etc.

5.3   Photo-Atlas

CSES usage needs some training, at least for those parameters whose rating is not sharply numerically delimited. Whereas *e.g.* Cliff slope, Beach width and Tidal range can be easily estimated, other parameters, *e.g.* Vegetation debris amount, Built environment quality and Skyline attractiveness assessment could be more debatable.

To assist surveyors, in the following Appendix photographs of each parameter/value is reported, which may be used as reference cases. Ratings for those sceneries have been assessed on the basis of the experience done during the last few years, when approximately 1000 locations were assessed in different countries/continents by the authors of this book.

5.4   Conclusions

Coastal areas of the world are under threat due to the forcing function of people who flock to the coast for habitation and/or recreation. This squeeze affects an extremely strategic asset – coastal scenery itself.

Coastal managers together with planners need coastal landscape inventories in order to base sound management decisions on ascertained facts. Today, most scenic assessments have been carried out on a subjective basis. Instructions to surveyors are necessary as well, since not all the parameters are numerically measurable and experience helps in obtaining their correct rating.

A Photo-Atlas showing all the 26 considered parameters (comprising physical, and human parameters) on which the CSES is based together with their ratings is provided. This gives surveyors an instrument in which to perform CSES via visual matching with landscapes evaluated by the authors of this book, as a first step in quantifying beautiful coastal scenery. These parameters were obtained by consultation with coastal experts and beach users. Each parameter was rated on a five-point scale, essentially covering presence/absence or poor quality (1), to excellent/outstanding (5).

Although parameter grading maintains a margin of uncertainty, data processing via fuzzy logic produces reliable results and the method has been proved to be robust.

The checklist presented and explained was designed to have a universal applicability with respect to coastal scenery. The goal of defining beauty may be illusory, but the pursuit of quantifying beauty is a worthwhile goal.

Appendix: Photo Atlas

	1 Cliff height (H)
	**1**: Absent (< 5 m)
	**2**: 5 m $\leq H$ < 30 m
	**3**: 30 m $\leq H$ < 60 m
	**4**: 60 m $\leq H$ < 90 m
	**5**: $H \geq$ 90 m

(continued)

	2 Cliff slope
	1: < 45°
	2: 45°–60°
	3: 60°–75°
	4: 75°–85°
	5: Circa vertical

(continued)

	3 Cliff special features
	1: Absent
	2: 1 special feature (e.g. rock fall)
	3: 2 special features (e.g. banding and terracing)
	4: 3 special features (e.g. banding, jutting cliff, rock fall)
	5: Many >3 special features (e.g., irregular profile, notch, banding, folding)

(continued)

4 Beach type
1: Absent
2: Mud
3: Cobble/boulder
4: Pebble/gravel
5: Sand

(continued)

	5 Beach face width (W)
	1: Absent
	2: < 5 m or > 100 m
	3: 5 m ≤ W < 25 m
	4: 25 m ≤ W < 50 m
	5: 50 m ≤ W ≤ 100 m

(continued)

	6 Beach face colour
	1: Absent
	2: Dark
	3: Dark tan
	4: Light tan/bleached
	5: White/gold

(continued)

	**7 Rocky shore slope**
	1: Absent
	2: < 5°
	3: 5°–10°
	4: 10°–20°
	5: > 20°

(continued)

	8 Rocky shore extent
	1: Absent
	2: < 5 m
	3: 5 m – 10 m
	4: 10 m – 20 m
	5: > 20 m

(continued)

	9 Rocky shore roughness
	1: Absent
	2: Distintly jagged
	3: Deeply pitted and/or irregular (uneven)
	4: Shallow pitted
	5: Smooth

(continued)

	10 Dunes
	1: Absent
	2: Remnants
	3: Fore-dune
	4: Secondary ridge
	5: Several

(continued)

	11 Valley
	1: Absent
	2: Dry
	3: Stream (< 1 m)
	4: Stream (1 m – 4 m)
	5: > 4 m

(continued)

	12 Skyline landforms
	1: Not visible
	2: Flat
	3: Undulating
	4: Highly undulating
	5: Mountainous

(continued)

	**13 Tide**
	1: Macro (> 4 m)
	3: Meso (2 m – 4 m)
	5: Micro (< 2 m)

(continued)

	14 Coastal landscape feature
	**1**: None
	**2**: 1 Feature (e.g. blowhole)
	**3**: 2 Features (stack and shoals)
	**4**: 3 Features (e.g., caves, pinnacles, emayment)
	**5**: > 3 Features (e.g. irregular headlands, stacks, islands, shoals)

(continued)

15 Vista

1: Open on one side

2: Open on two sides

4: Open on three sides

5: Open on four sides

(continued)

16 Water colour
1: Muddy brown/grey
2: Milky blue/green; opaque
3: Green/grey blue
4: Clear blue/dark blue
5: Very clear turquoise

(continued)

	**17 Natural vegetation cover** **1**: Bare (< 10% vegetation only)
	**2**: Scrub/Garigue/grass (marram/gorse/ferns, bramble/meadow, etc.)
	**3**: Wetlands/meadow
	**4**: Coppice, Maquis (mature trees bushes)
	**5**: Varity of mature trees/mature natural cover

(continued)

18 Vegetation debris

1: Continuous >50 cm high

2: Full strand line

3: Single accumulation

4: Few scattered items

5: None

(continued)

	19 Noise disturbance
	**1**: Intolerable (e.g. loud music from speakers)
	**2**: Tolerable (e.g. Crowded but no speakers)
	**4**: Little (e.g. not crowded and without speakers; small communication lines)
	**5**: None (e.g. few or scattered people without radios, etc.)

(continued)

	20 Litter
	1: Continuous accumulations
	2: Full strand line
	3: Single accumulation
	4: Few scattered items
	5: Virtually absent

(continued)

	**21 Sewage discharge** **1**: Sewage evident
	**3**: Some evidence
	**5**: No evidence of sewage

(continued)

	22 Non built environment
	**1**: None
	**3**: Hedgerow/terracing/ monoculture
	**5**: Field mixed cultivation +/− trees/natural

(continued)

	23 Built environment
	1: Heavy industry
	2: Heavy tourism and/or urban
	3: Light tourism and/or urban and/or sensitive industry
	4: Sensitive tourism and/or urban
	5: Historic and/or none

(continued)

	**24 Access**
	1: No buffer zone/heavy traffic
	**2**: Buffer zone/light traffic
	**4**: Parking lot visible from coastal area
	**5**: Parking lot not visible from coastal area

(continued)

	25 Skyline
	1: Very unattractive
	2: Unattractive
	3: Sensitively designed
	4: Very Sentitively designed
	5: Natural/historic features

(continued)

	26 Utilities
	**1**: > 3 Utilities (e.g. slipway, handrails, electricity poles, sewage outfall)
	**2**: 3 Utilities (e.g. groins, electricity poles, railway)
	**3**: 2 Utilities (e.g. groin, lifeguard tower)
	**4**: 1 Utility (e.g. tetrapods)
	**5**: None

Reference

Ergin A, Karaesmen E, Micallef A, Williams AT (2004) A new methodology for evaluating coastal scenery: fuzzy logic systems. Area 36:367–386

Chapter 6
Examples of Class Divisions and Country Synopsis for Coastal Scenic Evaluations

Giorgio Anfuso, Allan T. Williams, and Nelson Rangel-Buitrago

> *Progress is impossible without change and those who cannot change their minds cannot change anything.*
>
> *George Bernard Shaw*

Abstract In the last decade, the book authors classified by means of a Coastal Scenic Evaluation System (CSES) technique using a checklist of 18 physical and 8 anthropogenic parameters, 952 coastal sites located around the world. Each parameter was rated from 1 (absence/poor quality) to 5 (outstanding quality) and a "D" value obtained which allowed site categorization into five classes: Class I (extremely attractive natural sites) to Class V (very unattractive urban sites). Almost 50% of investigated sites belonged to Class III (n = 197) and IV (n = 248). Class V sites (n = 209) were also well represented, followed by Class I (n = 155) and II sites (n = 143). Examples of each class from different locations are presented to give the reader a solid idea of the characteristic and essence of each class. Classes I and II generally correspond to natural sites in (or close to) protected areas with low or little human impacts giving high scenic values. Classes III and IV usually correspond with sites of intermediate to low scenic value because of poor scores of both physical and anthropogenic parameters. Class V sites are usually greatly affected by human impact and at many places, are linked to emplacement of chaotic defence structures built to halt/slow down coastal erosion processes. At places, scenic evalu-

G. Anfuso (✉)
Departamento de Ciencias de la Tierra, Facultad de Ciencias del Mar y Ambientales, Universidad de Cádiz, Cádiz, Spain
e-mail: giorgio.anfuso@uca.es

A. T. Williams
Faculty of Architecture, Computing and Engineering, University of Wales, Trinity Saint David, Swansea, Wales, UK

CICA NOVA, Nova Universidade de Lisboa, Lisbon, Portugal

N. Rangel-Buitrago
Departamentos de Física y Biologia, Facultad de Ciencias Básicas, Universidad del Atlántico, Barranquilla, Atlántico, Colombia
e-mail: nelsonrangel@mail.uniatlantico.edu.co

© Springer International Publishing AG, part of Springer Nature 2019
N. Rangel-Buitrago (ed.), *Coastal Scenery*, Coastal Research Library 26,
https://doi.org/10.1007/978-3-319-78878-4_6

ation covered large areas, e.g. the Caribbean coast of Colombia, the Mediterranean coast of Morocco, the Andalusia region of Spain, Mediterranean Turkey and the whole coast of a country, e.g. Bonaire, Cuba and Wales. These are presented as case studies: the main physical characteristics, management issues and class distributions are described, allowing a complete overview on these areas.

6.1 Introduction

Il faut s'entraider: c'est la loi de la nature. (J. de la Fontaine, Fables VII 17, L'Ane et la chien)

Over a time span of more than a decade, the book authors have assessed 952 global coastal locations (Fig. 6.1 and Appendix 1) within the framework of tens of Master/ Ph.D. thesis and national/international projects, e.g. Ergin et al. (2004, 2006), Langley (2006), Ullah et al. (2010), Williams et al. (2012), Rangel-Buitrago et al. (2013), Anfuso et al. (2017).

At each location, an assessor puts a tick in the relevant box, ranging from a low (1) to high (5), of the 18 physical and 8 anthropogenic parameters (Chapter 5; Figure 5.1; Ergin et al. 2006). Based on summed parameter evaluations, a final "D" value is computed, which divides sites into five distinct classes, from Class I (extremely attractive natural sites) to Class V (very unattractive, intensively developed urban sites). The number of sites analysed in each country and their distribution per continent together with the total amount recorded in each of the five classes is presented in Figs. 6.1 and 6.2.

The Caribbean coast of Colombia showed the greatest number of sites and their distribution was almost constant along approximately 2445 km of coastline from Panama in the south to the Venezuela border in the northeast (Rangel-Buitrago et al. 2013, 2017, 2018; Williams et al. 2016). In southern Brazil, within the regions of Rio Grande do Sul and Santa Catarina 120 sites were investigated (Oliviera et al. 2016; Cristiano et al. 2016). Cuba coastal scenery was investigated at 101 sites spaced all around the island and cays (Anfuso et al. 2014, 2017). The Spanish coast was investigated at 98 sites distributed in Andalusia, the Basque Country and the Catalonian coast (Williams et al. 2012; Iglesias et al. 2017). Other countries having a relatively high number of investigated sites were Malta, Turkey, Italy, Poland, New Zealand and Croatia (Fig. 5.2a).

Most investigated sites were located in Latin America, essentially on the Caribbean Sea and South Atlantic coasts (i.e. Bonaire, Brazil, Colombia, Chile and Cuba, n = 492). Europe also had a good representation (n = 322), a lower distribution of sites was observed at other continents (Fig. 6.2b).

Classes I (n = 155) and II (n = 143) show very close numbers (Fig. 6.2c) and generally correspond to natural sites in (or close to) protected areas with low or null human impacts and high scenic values. Almost half of the investigated sites belonged to Classes III (n = 197) and IV (n = 248, Fig. 6.2c), which usually correspond with sites of intermediate to low scenic value having poor scores of both physical and

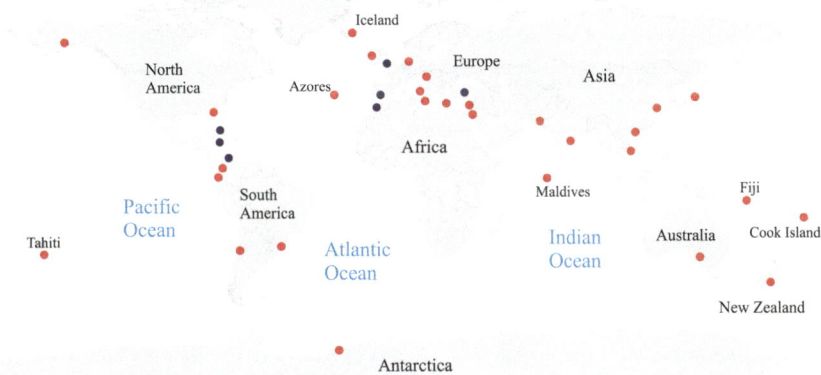

● Distribution of assessed sites: Antarctica, Australia, Brasil, Chile, China, China - Hong Kong, Cook Islands, Croatia, Cyprus, Ecuador, Egypt, Fiji, Iceland, India, Ireland, Italy, Italy - Eolian Islands, Japan, Jordan, Maldives, Malta, New Zealand, Pakistan, Peru, Poland, Portugal, Portugal- Azores, Tahiti, United Kingdom, Florida (USA) and Vietnam

● Large covered areas: Bonaire, Caribbean Colombia, Cuba, Andalusia (Spain), Mediterranean Morocco, Mediterranean Turkey and Wales

Fig. 6.1 Location map showing the distribution of countries with coastal sites assessed (n = 952)

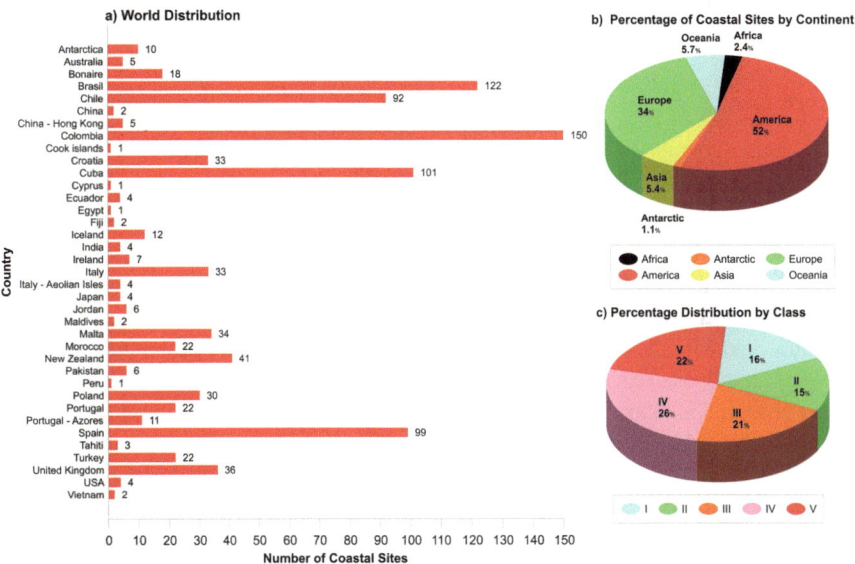

Fig. 6.2 (**a**) Number of the above selected sites analysed in each country (**b**) Percentage of coastal sites by continent (**c**) Percentage distribution by class

anthropogenic parameters. Class V, usually urbanised sites greatly affected by human intervention, are also well represented (n = 209, Fig. 6.2c).

The following Sect. 6.2 provides examples for each of the five classes from different global locations. This is to give the reader real examples of the characteristics of different classes, to understand the importance of each of the anthropogenic and physical parameters in class determination, as well as improvements that might be carried out to upgrade a site to a better class. Section 5.3 gives a representative coverage of the technique applied to larger areas, i.e. the Caribbean coast of Colombia, the Mediterranean coast of Morocco, the Andalusia region of Spain, Mediterranean Turkey and at national level, i.e. Bonaire, Cuba and Wales.

6.2 Analyses of Individual Locations As Examples of the Five Classes

'The time has come'. The walrus said, 'To talk of many things'. (Lewis Carol. Through the looking glass)

Analysis of Scenic evaluation Histograms, Membership degree and Weighted average curves for a small sample of the 952 investigated sites (Appendix 1) are presented below according to their class. The provided examples are not indicative of the entire coastal scenery of the country selected and the pictures provided reflect the location but not necessary all the scenic parameters.

6.2.1 Class I

This class invariably includes extremely beautiful natural sites with very high landscape values ($D \geq 0.85$). The excellent score is the result of very attractive natural features with very low or null level of human influence – indeed this is a basic requirement and usually Class I locations are located in Natural/National Parks and Protected areas, and consist of remote and rural sites (terminology according to Williams and Micallef 2009). The essence is that high attribute scores (4's, 5's; see Chaps. 4 and 5) are recorded.

As observed at many places (e.g. Spain: Williams et al. 2012; Colombia: Rangel-Buitrago et al. 2013; Cuba: Anfuso et al. 2014, 2017), topographic, geomorphologic and geologic settings acquire large importance. In the checklist (Table 5.1, Chap. 5), "cliffs" (points 1–3), "rocky shore" (points 7–9), "valley" (point 11)," skyline landform" (point 12) and (partially) "coastal landscape features" (point 14) are frequently linked to the presence of a high rocky coast and (often) an undulating or mountainous landscape. Rangel-Buitrago et al. (2013, 2017, 2018) from observations in the Caribbean coast of Colombia showed that good scores for Class I locations were mostly found in a natural protected area close to the Sierra Nevada de Santa Marta mountain chain (e.g. Arrecifes beach, see Colombian case study). A similar situation was observed by Williams et al. (2012) in Andalusia (e.g. Los Genoveces beach, see Spain case study) and by Anfuso et al. (2017) in Cuba (e.g. Playa Colorada beach, see Cuba study case), a remote mountainous site not located in a protected area.

Other sites with similar geomorphologic/geological settings and D scores were Poppit in Wales (Wales case study), Kamkoum El Baz on the Mediterranean coast of Morocco (Morocco case study) and 100 Steps in Bonaire (Bonaire case study). Class I locations can also be found along low coastlines, which show lower scores with respect to cliffs, rock shore, etc., e.g. cay Levisa beach in Cuba (Anfuso et al. 2017), or coral islands in the Maldives.

6.2.1.1 Reynisfjara, Iceland (D: 1.11)

Various agencies have responsibility for coastal management including local authorities for local planning matters, minor erosion problems, harbour development, etc. Umhverfisstofnun (The Environment Agency) and the national government have responsibilities regarding wider coastal management and nature conservation issues. There is a general right of access (right to roam policy) in Iceland. For example, Reynisfjara (Fig. 6.3a) has no protected status and is privately owned, but many thousands of visitors access the beach each year.

The Reynisfjara (Fig. 6.3b), black basalt sand beach with columnar basalt, which can also be seen in the next example of the Giant's Causeway in Northern Ireland, together with offshore stacks are a major tourist draw. Beach sediments are very mobile, varying in size from the occasional boulder but are mainly composed of large pebbles and coarse sand depending on tide, wind strength and wind direction. One of the authorities' main concerns is the danger of tourists being knocked over by powerful waves as they approach the water's edge and being dragged back into the sea. There have been many accidents and several fatalities here and nearby, most recently in January 2016 when a German tourist was drowned. Visitor access is much more heavily controlled due to the many accidents/several fatalities by rockfalls and drownings at the nearby Dyrhólaey nature reserve (Fig. 6.3c). The Environment Agency has now closed off sections of the beach for safety purposes as well as the disturbance to breeding wildlife. Apart from beach sediment colour, vegetation, no rocky shore or dunes, the site scores very highly on all other checklist parameters (Fig. 6.3a).

6.2.1.2 Giants' Causeway, Northern Ireland (D: 0.87)

It is a UNESCO World Heritage Site designated in 1986 and located in County Antrim on the north coast of Northern Ireland, some three miles (4.8 km) northeast of the town of Bushmills. It is a rocky coast landscape of Tertiary basalt (Palaeocene age, some 60 million years ago) cliffs and shore platforms, backed by boulder beaches and fine-grained storm swash deposits, extending along 3 km of coast with about a 0.5 km width and the location scored highly on all parameters (Fig. 6.4a). The Grand causeway is the largest of three rock outcrops which make up the Causeway.

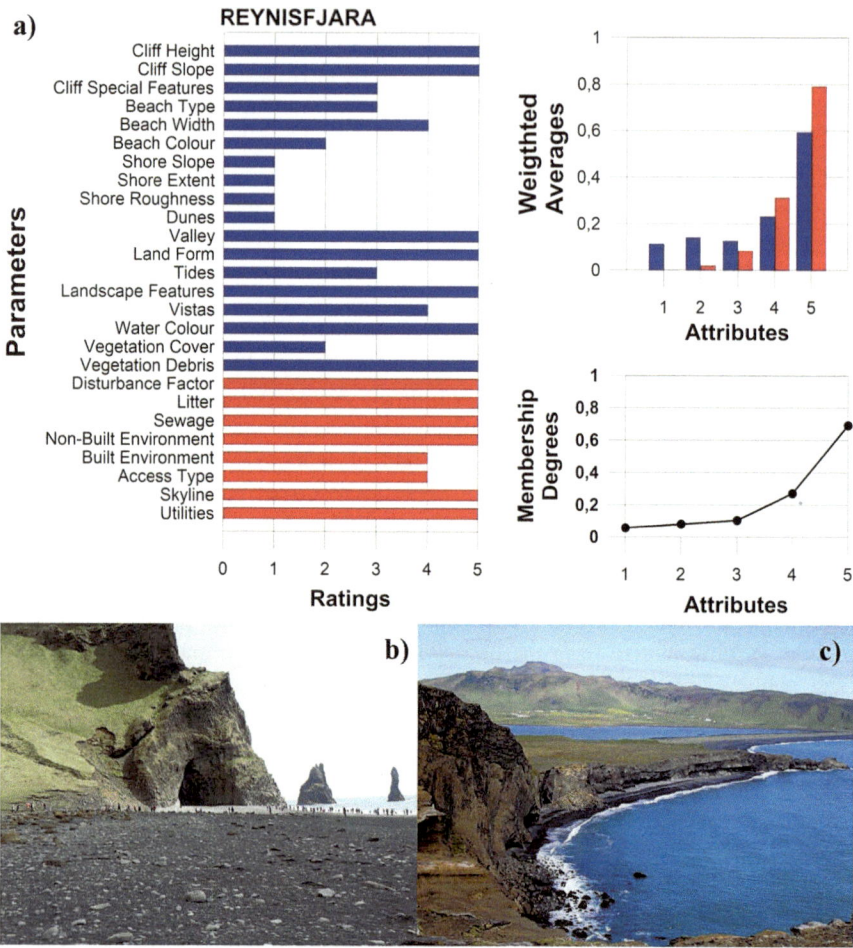

Fig. 6.3 (**a**) Scenic evaluation histogram, Weighted averages and Membership degree curve for Reynisfjara, Iceland (Class I, D: 1.11), (**b**) detailed view of Reynisfjara and (**c**) Dyrhólaey. (Photo by Andy Jones)

The shore platform centrepiece section is composed of typically hexagonal columns (12 m in height and ~40,000 in number formed by slow cooling when joints developed in the thick lava flow, Figs. 6.4b, c).

The high cliffs expose multiple basalt flows, some of which are also columnar, the tops often being marked by red lateritic soil horizons up to 10 m thick. Sub-horizontal basalt shore platforms developed backed by boulder beaches and fine-grained storm swash deposits. Many surface remnant boulders have been exhumed directly as rounded boulders from the weathered basalt, several still attached by pedestals to the underlying surface. In several peri-tidal rock pools contemporary

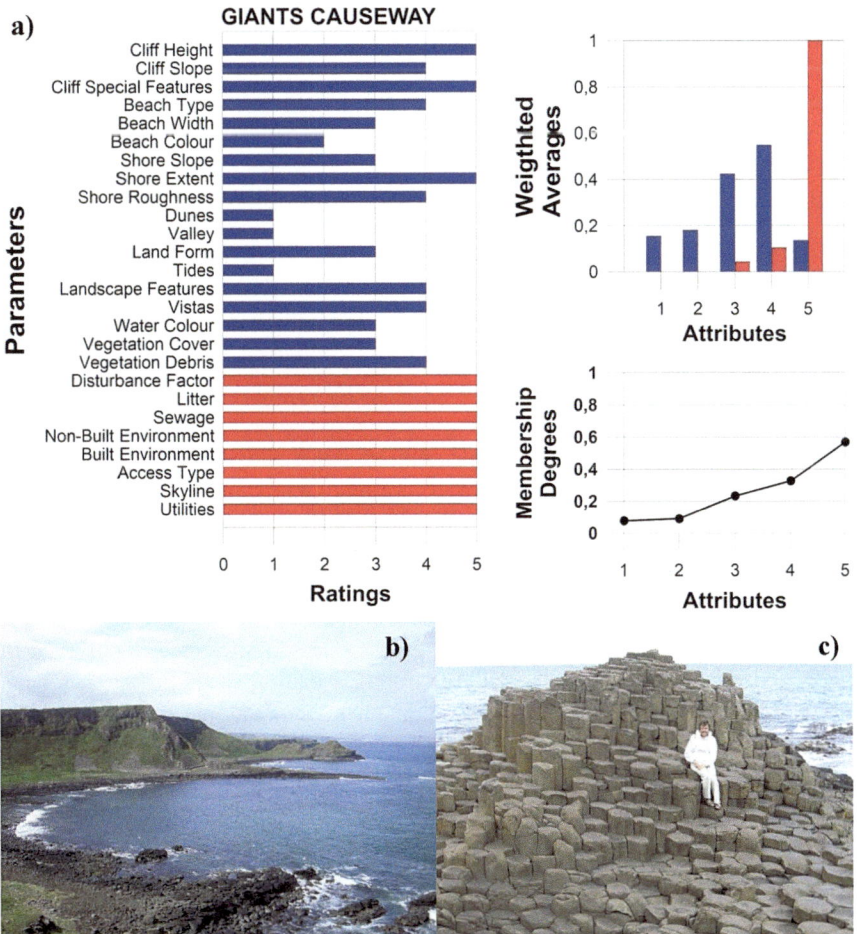

Fig. 6.4 Giants Causeway. (**a**) Scenic evaluation histogram, Weighted averages and Membership degree curve (Class I, D: 0.87), (**b**) Aerial view and (**c**) the basalt columns (Scale 1.50 m)

stromatolite formation is taking place. The cliff and human parameters are outstanding.

The site has a rich mythology associated with the various distinctive rock formations, e.g. Finn McCool the Irish giant was annoyed by the threat to Ireland by the Scottish giant Benandonner. He grabbed parts of the Antrim coast and threw them into the sea to form a path for him to follow and teach Benandonner a lesson. Wrong call!! Some rock formations are named, e.g. Chimney Stacks, the Giant's Boot (size 93.5 lost by Finn McCool as he fled from the wrath of Benandonner) and the Camel's Hump.

Access is via continuously monitored and maintained footpaths along the cliff top, a steep footpath and a paved road providing access to lower levels. Former paths have been closed due to instability and periodic slope movements but signs explaining the erosion processes are posted to increase visitors' awareness. A new visitor centre was opened July 12th 2012, at a cost of £18.5 million and the site is mostly under the ownership and management of the National Trust, in 2016 attracting >850,000 visitors.

Other Class I locations are given in Fig. 6.5, e.g. South Georgia (Figs. 6.5a–c), Australia (Figs. 6.5d, e) and the Maldives (Figs. 6.5f, g).

6.2.2 Class II

Coastal areas with D values $0.85 > D \geq 0.65$ usually include natural or semi-natural/urban sites with high landscape values, mainly close to protected areas with a low intrusion of human impact (acceptable anthropogenic activities/structures). These sites rate lower than Class I due to a lower scoring of landscape features, i.e. they do not present special features, cliff or rock shore but usually have high scores with respect to water and sediment parameters, dunes and well developed vegetation cover, or the good scores linked to natural parameters are slightly lowered because of human influence.

At very limited places (Anfuso et al. 2014), Class II sites can be observed in resort and urban areas, e.g. at Varadero (Cuba), a natural sand spit that has been urbanised since the 1970s for the development of international tourism, e.g. the beach in front of the Council Museum, as well as at many attractive tourist destinations in Cuban cays, e.g. Playa Pilar, cay Guillermo, which presents white, fine-grained sand beaches composed of coral fragments from adjacent reefs, turquoise water colour, luxuriant vegetation and well developed dunes. These are good example of appropriate coastal management to reach a Class II site: very good scores can be linked to an important change in the approach to coastal erosion and environmental problems, such as, those related to human usage of the back beach area and dunes.

The modalities adopted permit recovery by means of beach nourishment works, which favour large beach formations, dune reconstruction and vegetation recovery, which directly influences physical parameters and indirectly, several human parameters. A well vegetated dune works as a buffer area between beach and the backing urbanised areas, as it reduces noise disturbance linked to urban traffic and increases the skyline quality constituting a natural protection for any negative visual impact of the built environment.

Examples of beaches belonging to this class are Gore Bay. New Zealand and Baía do Sueste, Brazil:

Fig. 6.5 (**a–c**) Gritviken (D: 0.87), South Georgia. Background and old historic whaling tanks, (**d–e**) Dee Why (D: 0.85), Australia, a great surfing beach (photos by Andy Short), (**f–g**) Paradise Island (D: 1.10), Maldives, beach scenery on a low relief island

6.2.2.1   Gore Bay, New Zealand (D: 0.66)

Gore Bay, located near to Christchurch on the east coast of New Zealand, is a popular holiday destination, characterised by a wide beach and houses nestling in the native bush of the coastal hinterland. Housing is sensitively designed so that from the shore little is visible and good scores are observed for the natural parameters (Fig. 6.6a). It has been a small holiday resort for some time and is not particularly tourism focused leading to a probable lack of interest in locals modifying the environment.

Reasonable attribute scores occur for cliff, beach and the bush vegetation (Figs. 6.6b, c) and human parameters are the main reason for its Class II status (Fig. 6.6a). Little can be done with regard to the physical parameters apart from cleaning of vegetation debris. Beach access and utilities could be better and a buffer superimposed between beach and road, but this is unlikely to be carried out.

Weighted averages (Fig. 6.6a) bear out this point suggesting that the lack of human interference helps to emphasise the natural quality of the bay. Management of the area comes under the auspices of the 1991 Resource Management Act (RMA) which provides for the sustainable development of New Zealand's natural and physical resources. This legislation replaced 54 separate statutes. The RMA grants regional councils the authority to decide criteria outlining what should be considered as 'scenic value' yet nowhere in the plan is there a definition of 'outstanding landscape', which makes consistent decision making difficult. New Zealand' scenic assessment is wide ranging and subjective, which does not help reliable management approaches.

6.2.2.2   Baía Do Sueste, Brazil (D: 0.71)

Located on an island in a sheltered bay fronted by other small islands all belonging to the Archipelago of Fernando de Noronha, Northeast Atlantic Ocean of Brazil, the Archipelago is a UNESCO World Heritage Site with a delicate ecosystem. The area is part of PARNAMAR, a protected area that can be visited by a restricted number of users. The Baía do Sueste is an attraction due to the great number of sharks and sea turtles that can be found. Natural parameters have very good scores (Fig. 6.7a) and is a site with the most important geo-diversity of the island, as it includes all the geological formations of the archipelago (volcanic rocks of the Remedios and Quixaba formations and sedimentary formations of the Caracas Formation, Fig. 6.7b). Vegetation cover is luxuriant but the classification is lowered because of the presence of vegetation debris (Fig. 6.7b), which forms a distinct strand line and gives rise to unpleasant odours. Dunes have dry herbaceous/shrub vegetation within them (Fig. 6.7c) together with mangroves, the latter being the only place in which they occur in these oceanic islands.

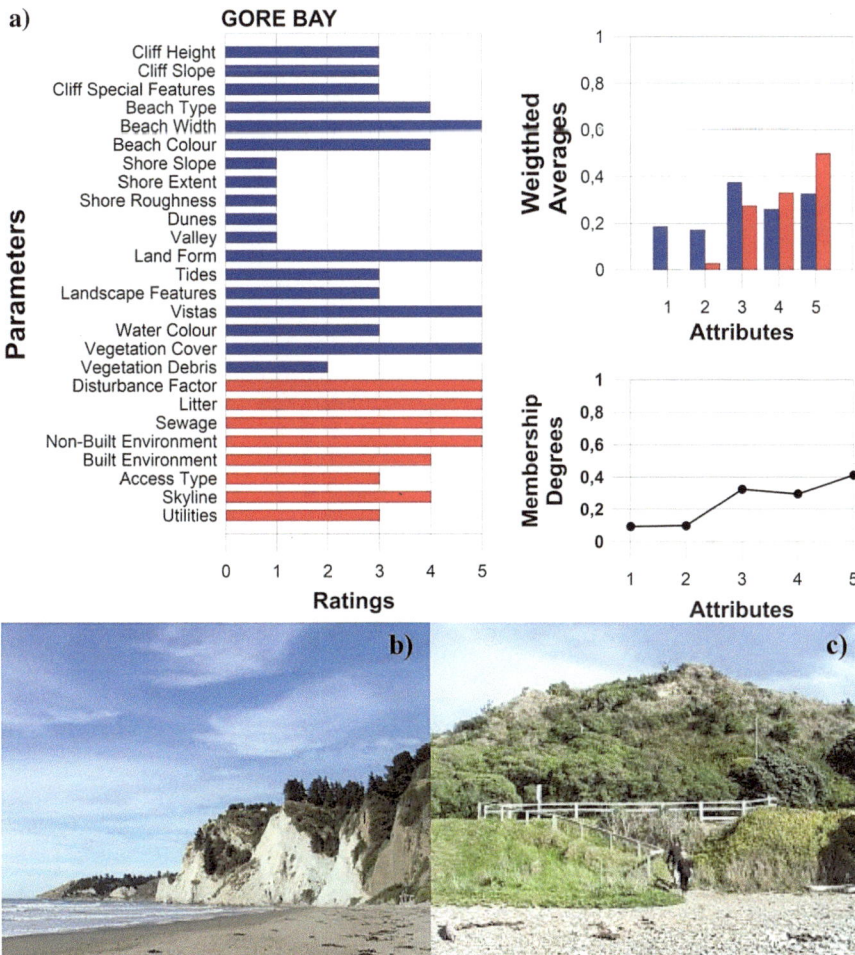

Fig. 6.6 (a) Scenic evaluation histogram, Weighted averages and Membership degree curve for Gore Bay, New Zealand (D: 0.66); (b) View of the beach and cliff and (c) beach access straight from road. (Photos by Richard Langley)

The beach does not record many visitors even during holiday periods because of its location within the PANAMAR, and disturbance factor, litter and sewage evidence have good scores. Skyline and utilities are negatively affected by the presence of an Information and Control Point with a snack bar, shop, bathroom, shower, diving equipment rental and parking lot, all visible from the coast.

Other Class II sites that conform to this classification can be found at Varadero Beach and Playa Pilar, Cuba (Figs. 6.8a, b), the pocket beach of Mwnt Bay, Wales (Fig. 6.8c) where despite high scores for the anthropogenic parameters, low scores were obtained for, e.g. vistas, absence of dunes and appreciable amounts of litter,

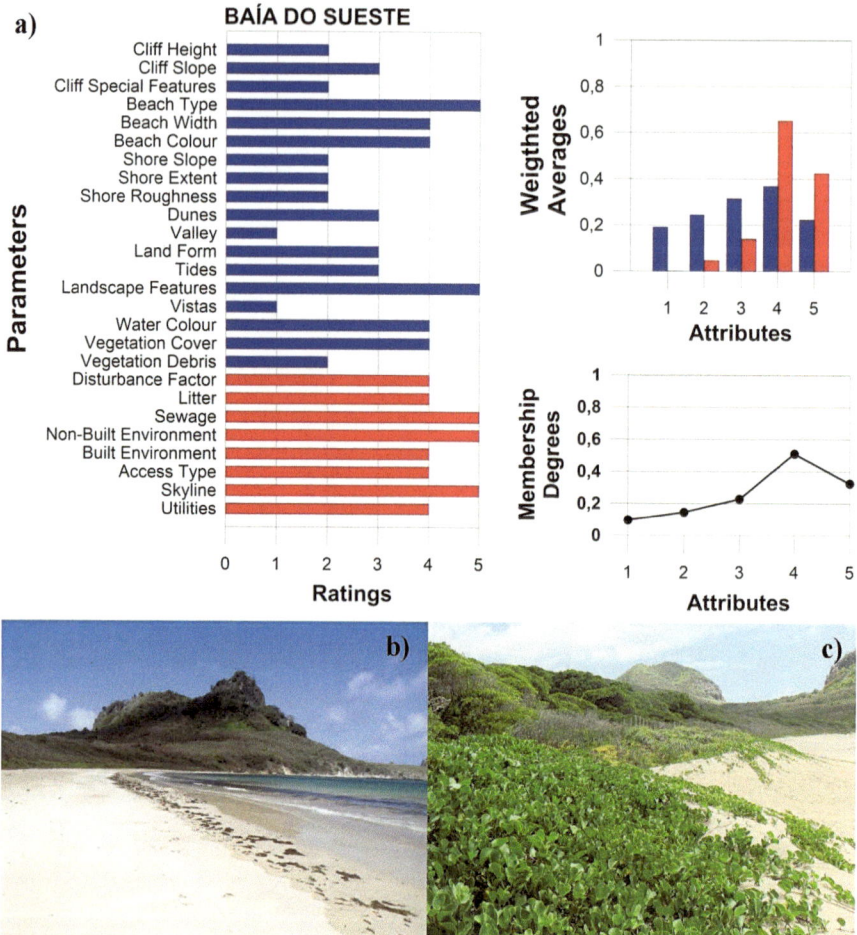

Fig. 6.7 (a) Scenic evaluation histogram, Weighted averages and Membership degree curve for Baía do Sueste, Brazil (D: 0.71) (b) General view of the area and (c) detailed view of dunes

and Plage Rouge, Morocco (Figs. 6.8d, e), negatively influenced by a massive presence of litter.

6.2.3   Class III

D values lie between 0.65 > D ≥ 0.40 and sites are usually village and some resort areas that present an attractive scenario, but one that is commonly flawed by features such as non-attractive buildings, no buffer zones, presence of litter etc. e.g. Caucana, Italy. This has been observed at Playa Blanca (see Colombia case study), Playa Ensenacho (see Cuba case study) and Conil (see Spain case study).

Fig. 6.8 (**a**) Varadero beach (D: 0.79) in front of the Council Museum, Varadero, Cuba; (**b**) Playa Pilar (D: 0.76), Cay Guillermo, Cuba; (**c**) Mwnt Bay (D: 0.72), Wales and (**d–e**) Plage Rouge (D: 0.66), Morocco

Examples of sites belonging to this class are Caucana, Italy and Goa Old Fort, India.

6.2.3.1   Caucana, Italy (D: 0.58)

This south Sicily site is close to an archaeological area of Byzantine age coastal settlements and today shows condominiums and summer houses essentially frequented only during July and August. The area has recorded a huge urban increase since the 1970s – especially the 1980s and 1990s and the hinterland is now totally urbanised with many coastal villages joining to form unique linear seaside resorts.

With regard to natural parameters, the beach along with dunes, obtains good scores (Fig. 6.9a); sediments being composed of fine golden quartz-rich sand (Fig. 6.9b). Dunes are vegetated by pioneer plants and Mediterranean bushes. At places, mature trees, essentially belonging to the Acacia family have been planted and form with dunes a buffer partially limiting the view to the buildings. A partially

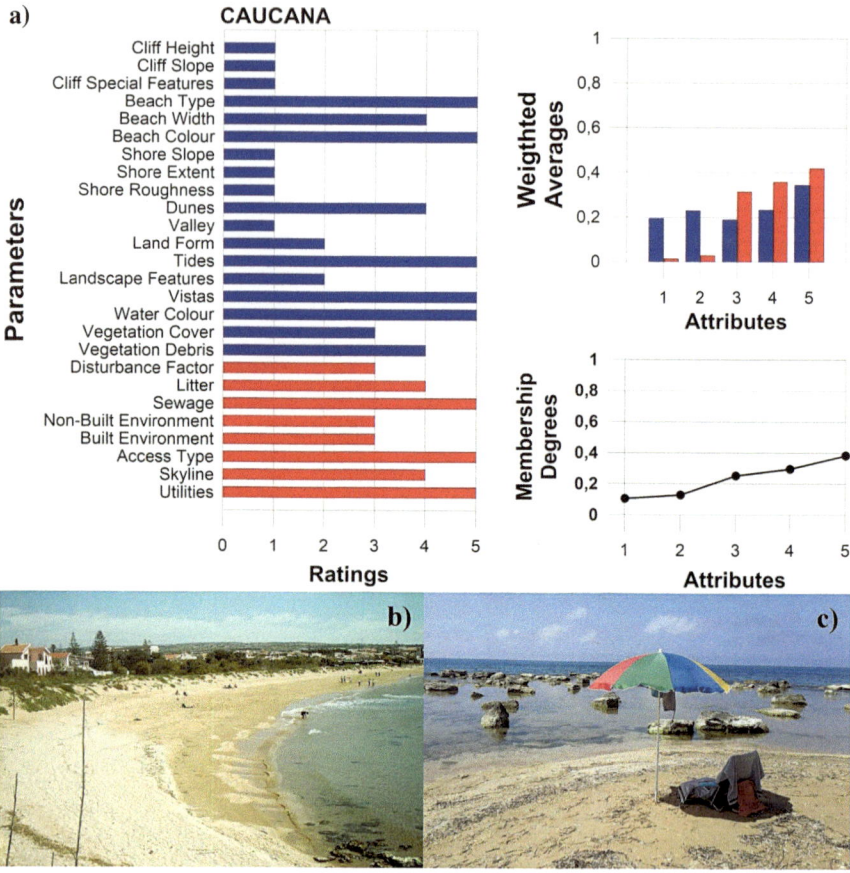

Fig. 6.9 (a) Scenic evaluation histogram, Weighted averages and Membership degree curve for Caucana, Italy (D: 0.58); (b) View of the area and (c) detail of a partially submerged breakwater limiting the beach westward

submerged breakwater limits the beach westward and has a low scenic impact because it was built in accord with a natural headland (Fig. 6.9c).

The beach is a family oriented environment and noise disturbance is not usually excessive, and beach sports, e.g. volley ball, football are played at some sectors of the beach. Litter presence is reduced because of the existence of many bins at principal beach access points and mechanical and manual cleaning operations are carried out with an almost daily periodicity.

6.2.3.2 Goa Old Fort, India (D: 0.55)

In the 1970s Goa's coastline was pristine and then a tourist boom commenced with hotels, bars, etc. proliferating along the coast reduced scenic scores of both natural and human parameters (Fig. 6.10a). Many past reports dealt with coastal management policies, but none were made public and many amendments made them obsolete. In

1991, the national government introduced the Coastal Regulation Zone, which became the focus of all coastal legislation, containing four categories, of which areas of outstanding natural beauty/historical heritage areas was one and all coastal states had to produce management plans. Since then, a National Coastal Zone Management Authority has been set up to promote integrated plans within all coastal states.

The nineteenth Century Portuguese fort (Fig. 6.10b) is a big tourist attraction but serious erosion is taking place and parts of the adjacent beach area are being eroded by some 5 m/year. Litter is a big problem (Fig. 6.10c) and solving this would help this site obtain a better scenic score. Other human parameters (Fig. 6.10a) could be slightly improved but this would still not make it a Class II location, as physical factors militate against this.

Other Class III sites are Austenmeer, Australia (Figs. 6.11a, b), a large number of utilities, car park right on the beach, houses not landscaped, render this site a Class

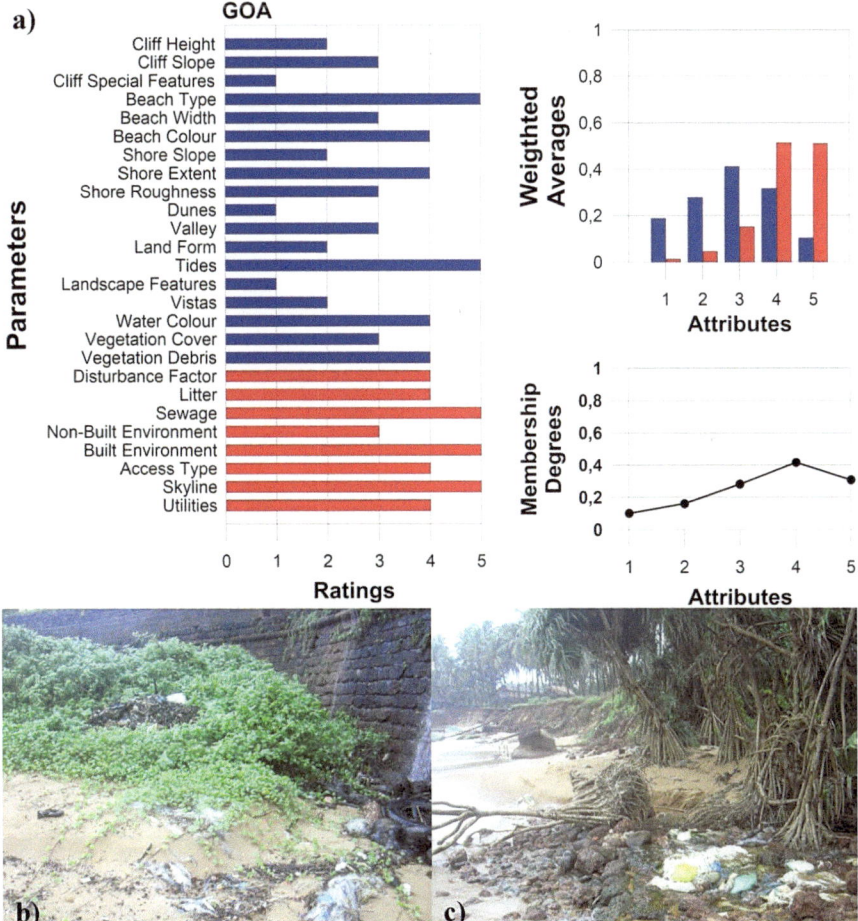

Fig. 6.10 (a) Scenic evaluation histogram, Weighted averages and Membership degree curve for Goa, India (D: 0.55); (b) Wall at the Old Fort and (c) beach litter in an adjacent area

Fig. 6.11 (**a–b**) Austenmeer, Australia (D: 0.62), with pools for bathing emplaced on rock shore platform (photos by Noel Kane –Maguire); (**c**) Praia do Cardoso Sul, Santa Catarina, Brazil (D: 0.42); (**d**) Santa Lucía, Camagüey, Cuba (D: 0.58); (**e**) Andrea Beach, Bonaire (D: 0.52).

III. With a slight effort, it could easily be a Class II. Praia do Cardoso Sul, Brazil (Fig. 6.11c, winter time), is a very attractive site that during holiday period has cars parked in the strand and has abundant litter items. Santa Lucia, Camagüey, Cuba (Fig. 6.11d) and Andrea Beach, Bonaire (Fig. 6.11e) are natural rural areas greatly affected by vegetation debris and litter.

6.2.4   Class IV

This class includes surveyed sites with 0.40 > D ≥ 0.00 values. They are mainly sites with poor landscape values significantly damaged due to undesirable anthropogenic activities. Examples of sites belonging to this class are South Beach-Public, Jordan and No 1 Beach, Qingdao, China.

6.2.4.1   South Public Beach, Jordan (D: 0.22)

The beach is in the middle of the town and scenically there is little that authorities can do to the bulk of the parameters (Fig. 6.12a).

Recreational boats (Fig. 6.12b) are present and affect noise disturbance and bathing safety. Beach debris can be cleaned but the reality is that any effort would only be superficial because of the presence of an unattractive skyline dominated by high buildings (Fig. 6.12c). Jordan has an extremely short coastline –some 27 km and Aqaba (population 100,000) is the coastal centre and coastal management is in its infancy. Many Action Plans have been invoked e.g. 1993, 2005 and 2012, but not until the Aqaba Special Economic Zone Authority was created that any semblance

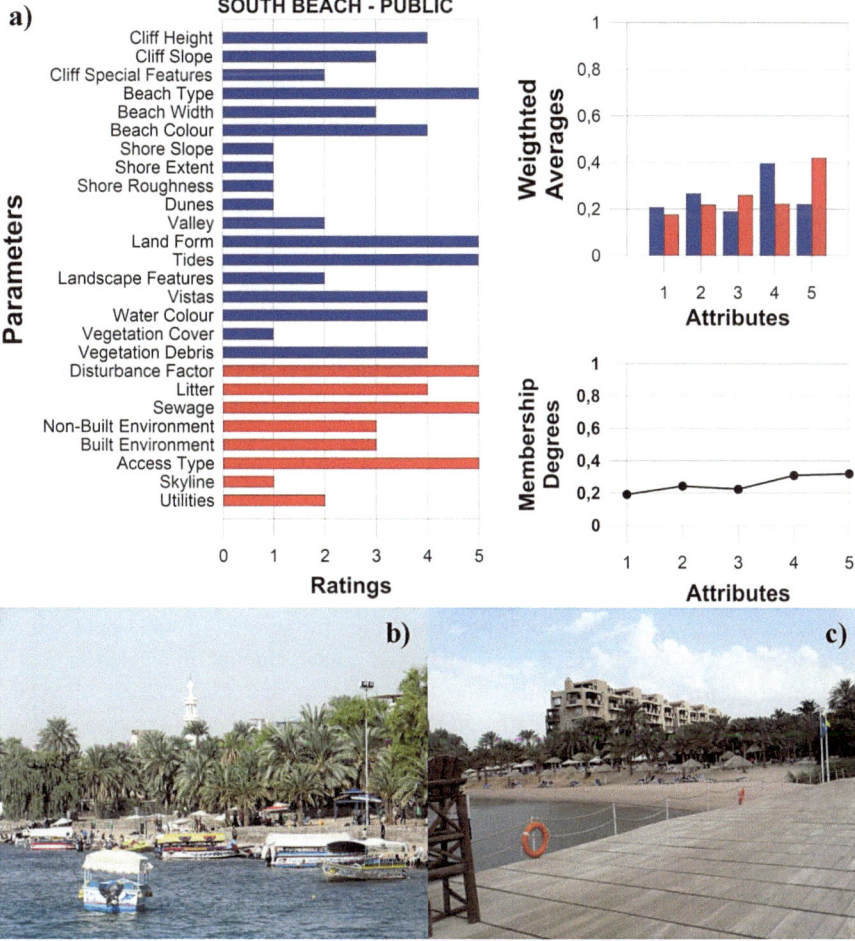

Fig. 6.12 (a) Scenic evaluation histogram, Weighted averages and Membership degree curve for South Beach, Jordan (Class IV, D: 0.22); (b) recreational boats in front of the beach and (c) nearby high rise hotels

of management was installed. This created a Marine Protected Area basically for coral reef management and it was a new paradigm for the country. A basic knowledge and understanding of Marine Spatial Planning exists but south of Aqaba and pre the port area development, a desert coast existed, in which a few tourist resorts have been built, e.g. Tala Bay.

6.2.4.2   N°1 Beach, Qingdao, China (D: 0.36)

The coastal area covering 3.6 ha, has been a favourite spot for tourists/locals for more than 100 years, and more than 1 million beach users per year have been counted. It has undergone many alterations in its history, the latest being after the founding of New China (1949), and since then local government has restored it many times, especially after the 2008 Olympic Games sailing events were held nearby.

In 1979 enactment of "coastal zone management regulations" was realised, but no laws and administrative regulations issued. Specific legislation for all types of capital source and regional expertise exists, but relevant laws for the coastal zone as a system do not exist at the time of writing. Local legislation is very few, only Jiangsu Province and Qingdao City have developed their own Local laws and regulations. In the coastal zone, there are some 20 China sea-related departments, such as: the Ministry of Agriculture and Fisheries; Traffic Management departments, etc. As can be seen from Fig. 6.13a, only on landscape features and sewage does the beach score highly with respect to scenery. Little can be done for either the physical or anthropogenic parameters (Fig. 6.13b) for this beach, as scenically it is poor, but it remains a fine recreational beach, although this has been marred somewhat by green tide (Fig. 6.13c, *Enteromorpha prolifera*) invasions especially between 2007 and 2010. A rapid expansion of *Porphyra yezoensis* aquaculture in Jiangsu province, which provided a greatly enhanced source of *E. prolifera*, along with global warming, seems a likely cause of the invasion.

Other areas that conform to this class are Repulse Bay, China (Fig. 6.14a), which has very poor human parameters; Magellan Foreland, Ireland (Fig. 6.14b), which has good human parameter scores but low scores for several physical parameters; Bocatocinos, Colombia (Fig. 6.14c), which has a huge amount of vegetation debris and litter items supplied by the Magdalena River; and Coco Beach, Bonaire (Fig. 6.14d), which records negative values of human parameters because of utilities, litter, etc.

6.2.5   *Class V*

These are locations with D < 0.00 and sites have very low scores of both natural and human parameters. However, the usual factor causing a low D value is human impact. In the Makadi example given below, it is the skyline, utilities and non-built

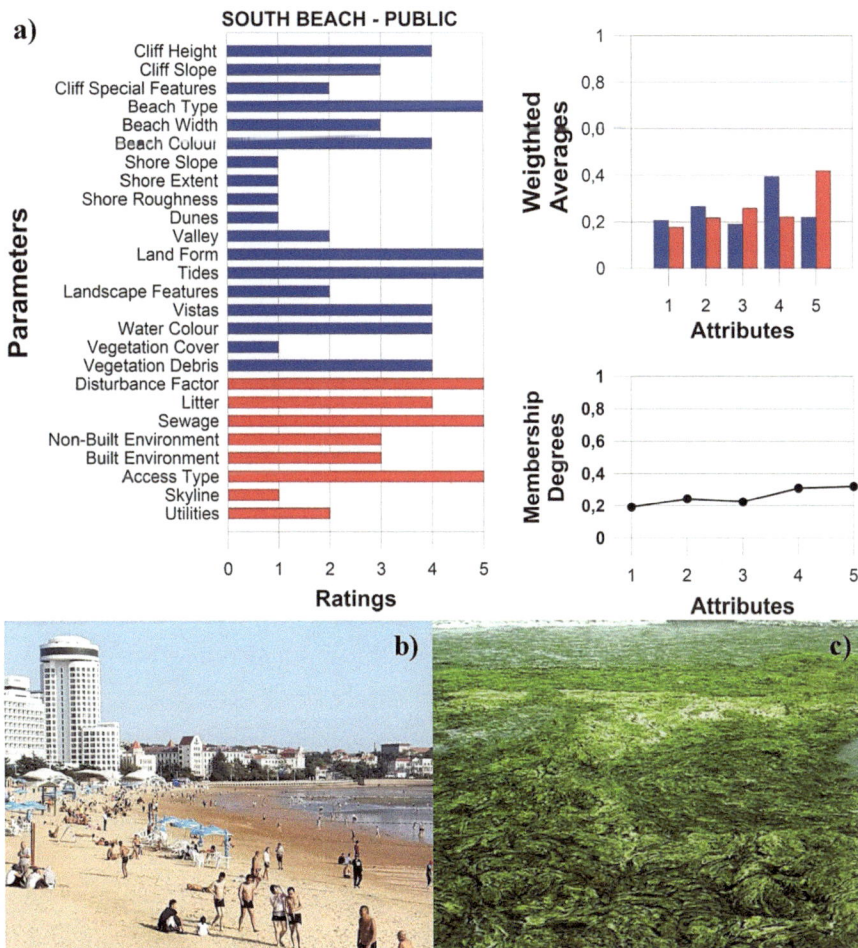

Fig. 6.13 (a) Scenic evaluation histogram, Weighted averages and Membership degree curve for No 1 Beach (D: 0.36), China; (b) High buildings and (c) view of green tide. (Photos by Hong Bin Liu)

environment that bring the attribute values down considerably. Sites belonging to this class are Makadi, Egypt and Playa Tarqui, Ecuador.

6.2.5.1   Makadi, Hurgahada, Egypy (D: -0.02)

On the Egyptian Red Sea coast, coastal management is purely a function of the many various resorts, usually owned and run by international companies that are dotted about the coastal region. Beyond the confines of these resorts there exists simply desert where litter tends to accumulate along locations and non-resort hinterland

Fig. 6.14 (**a**) Repulse Bay, China (D: 0.04); (**b**) Magellan Foreland, Ireland (D: 0.30); (**c**) Bocatocinos, Colombia (D: 0.05) and (**d**) Coco Beach, Bonaire (D: 0.22)

areas. This is a region that is rarely cleaned and litter eventually is carried far into the desert. An example of the above class is Makadi, Hurgada (D: -0.02, Fig. 6.15).

The resorts are planted with trees, etc. and buildings are usually low bungalow style in type and exist as oases of calm in a barren desert landscape. Similar scenarios are exemplified also in the Tala Bay resort in Jordan (see Class IV, Jordan). In spite of this, a glance at Fig. 6.15 will show that the area could be considerably improved. The beach is composed of dead coral and has a soft white/golden colour, which in conjunction with clear water makes it very attractive for tourists (Figs. 6.15a, b). The coral comes from a reef that extends seawards some 20 m offshore. Any litter blown into the area is immediately cleaned away by resort staff. Outside the resort

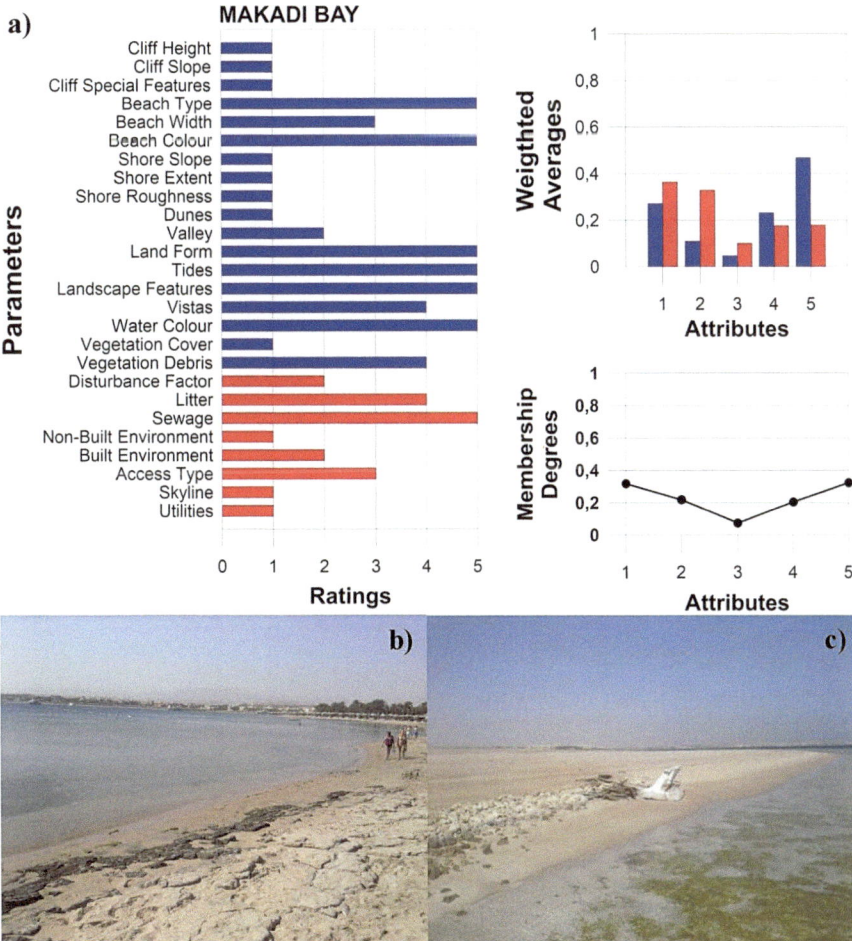

Fig. 6.15 (a) Scenic evaluation histogram, Weighted averages and Membership degree curve for Makadi, Egypt (D: -0.02, photos by Viv Griffiths); (b) The beach sector shows good water quality and no litter/seaweed (manual cleaning); and (c) view outside the resort area

area no cleaning operations are carried out and algae, litter and, at many places, remnants of concrete structures, can be observed (Fig. 6.15c). Within the resort, communication towers and pylons all help to lower anthropogenic parameters.

6.2.5.2 Playa Tarqui, Ecuador (D: -0.41)

The site has an approximate length of 900 m and constitutes a sector of Manta urban beach, on the Pacific Ocean. Apart from beach characteristics, vistas and water colour, human parameters have poor scores (Fig. 6.16a), e.g. cars parked on the beach (Fig. 6.16b), high rise buildings (Fig. 6.16c).

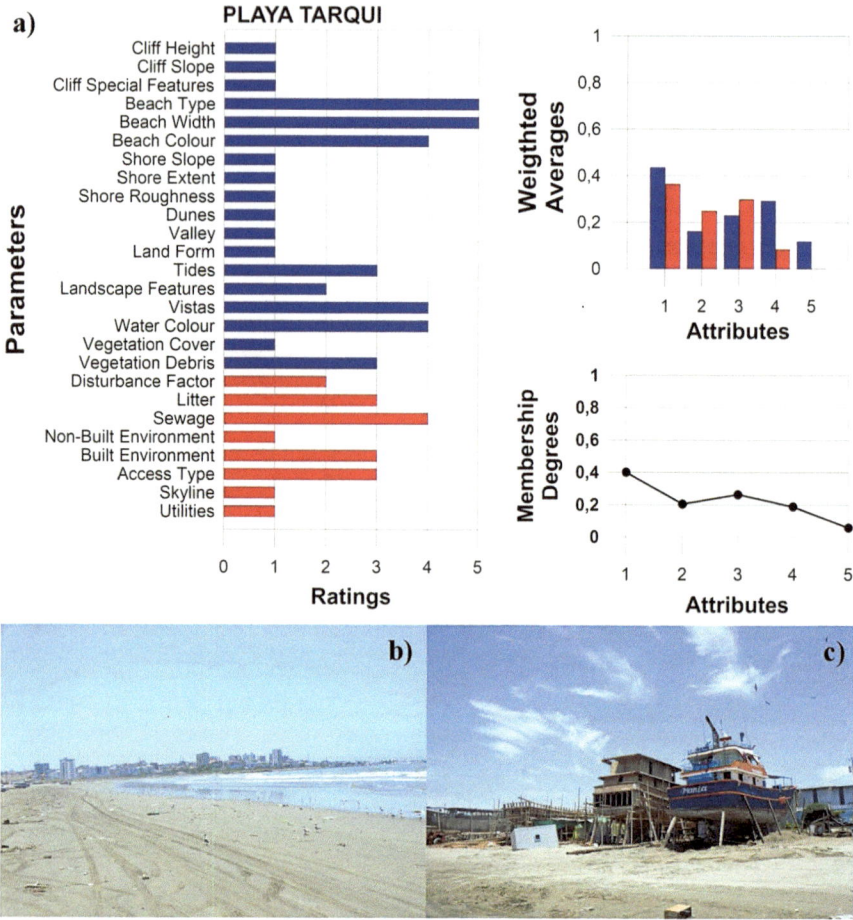

Fig. 6.16 (a) Scenic evaluation histogram, Weighted averages and Membership degree curve for Playa Tarqui, Manabí, Ecuador (D: -0.41); (b) cars are parked in the upper part of foreshore and built environment and skyline parameters are greatly affected because of the presence of high buildings on the western part of the beach and (c) view of boats under construction

A fish and seafood market is held every early morning on the east end of this beach, with vendors selling shark, swordfish, tuna, dolphin fish, octopus, scrimps, mussels and, according to locals, probably one of the best 'ceviche' in the country.

Along Class V sites, natural processes, such as, coastal erosion are often linked to inappropriate human interventions (e.g. interruption of longshore drift, dam construction, etc.), which produces beach width reduction/disappearance, an increase of sediments or of grain size (from sand to pebble and/or cobble) and, eventually, dune erosion (Fig. 6.17).

Human response, such as, emplacement of disordered defence structures has had an even more negative effects on scenery (Rangel-Buitrago et al. 2013, 2017, 2018) along many sectors of the Caribbean coast of Colombia, e.g. Galerazamba, Colombia

Fig. 6.17 (**a**) Tierrabomba, Colombia (D: -0.58); (**b**) Morales village, Holguin, Cuba (D: -0.09); (**c**) Playa Piedra Larga, Manabí, Ecuador (D: -0.99), the shoreline is heavily armoured and (**d**) Arroio do Silva, Brazil (D: -0.59), with evidence of sewage discarge

case study, and Tierrabomba (Fig. 6.17a), and Anfuso et al. (2014) in the southern coast of Cuba, e.g. Mayabeque, Cuba case study, and Morales (Fig. 6.17b). A similar situation can be observed at Playa Piedra Larga, Ecuador (Fig. 6.17c), where the dry beach is very narrow and the shoreline is protected by a revetment. Another location which fits into this class is Arroio do Silva, Santa Catarina, Brazil (Fig. 6.17d), whose natural parameters are greatly affected by human impacts.

Many Class V beaches are observed at the degraded areas along the Caribbean coast of Colombia and Cuba (see case studies), but also at other urban areas all around the world, where chaotic coastal defence structures have been emplaced to stop coastal erosion and/or at places where human occupation has deeply trans-formed the landscape, e.g. the 'concretisation' of the Spanish Costa del Sol.

6.3 Scenery Evaluation Applied at Large Areas/National Level

What is the use of a book, 'thought Alice,' without pictures or conversation. (Lewis Carol, Alice in Wonderland, ch 1)

At some areas, scenic evaluation was carried out at nearly homogeneously spaced sites, which allowed a representative cover of large areas, i.e. the Caribbean coast of Colombia, the Mediterranean coast of Morocco, Andalusia, Basque Country and Catalonia regions in Spain, and Mediterranean Turkey and samples from the whole coast of a country (i.e. Bonaire, Cuba and Wales). Below is a general description of investigated sites presented as independent case studies (Fig. 6.18).

As previously mentioned in Sect. 5.1, Class I sites are usually not well represented. An exception is observed at Bonaire where more than 50% of sites belong to this class (Fig. 6.18). Such good scores are linked to the existence of many protected natural areas of great scenic interest; c. 30% of the total country land area has been protected since 1969. Class II shows similar numbers to Class I with numerous cases observed at other places, i.e. at cays in Cuba, and in Wales. Classes III and IV are usually the most represented and at many places, constitute more than 50% of all classes (i.e. Cuba, the Mediterranean coast of Morocco, Andalusia, Basque Country and Catalonia regions in Spain and Wales, Fig. 6.18). Class V sites usually reflect the direct/indirect influence of human interventions, e.g. the presence of buildings, protection structures, litter, noise, etc., and are quite abundant at all investigated areas (Fig. 6.18).

The following examples are not indicative of the entire coast of the countries selected and the pictures provided reflect the location but not necessary all the scenic parameters.

6.3.1 Bonaire

Bonaire is the most easterly island in the Old Dutch Caribbean region. It is crescent shaped and oriented NW-SE, approximately 40 km long by 11 km at its widest point, with a land area of 288 km^2 where 18,905 inhabitants are found mainly in two towns: Kralendijk (the capital) and Rincon. At present, cruise, beach and dive tourism are the most important economic activities of the island, and the capacity for growth is rapidly increasing due to the warm, dry climate and attractive natural environment.

Since 1969, nearly 30% of the total land area has been protected as a national park. The Bonaire National Marine Park (BNMP) was created in 1979 and has had consistent management since 1991. It includes all the waters surrounding Bonaire (since 2001), from the high-tide mark to 60 m depth. It has many beaches and 2700 hectares of coral reef, seagrass and mangrove ecosystems, which provide habitat for a diverse range of marine species including about 65 species of stony coral and

Study Cases - Percentage Distribution by Class

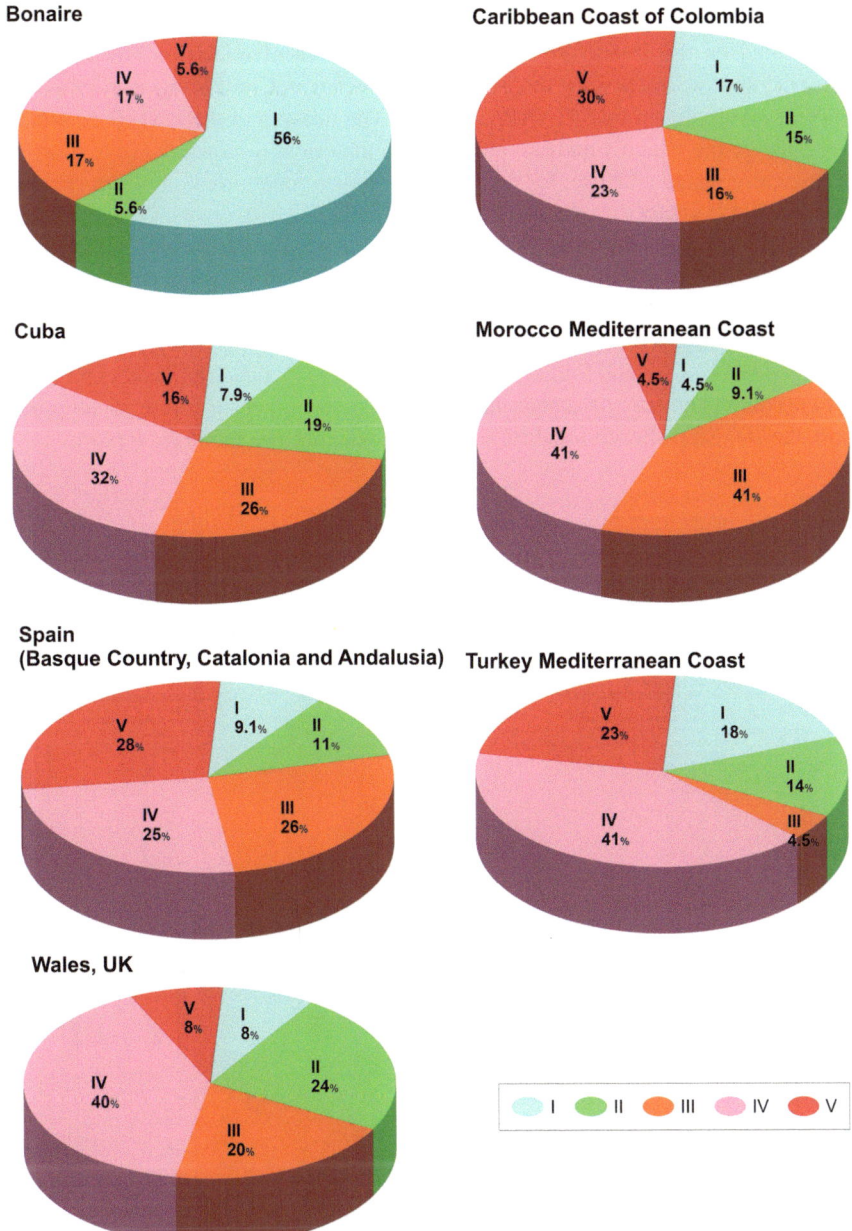

Fig. 6.18 Class distribution along each of the investigated areas

more than 450 reef fish species. The park is managed by a local non-governmental, not for profit organization, STINAPA Bonaire, which has a co-management structure with stakeholders, conservationists and local interest groups, all of which are represented on the Board. The day to day management is carried out under the supervision of a Director, by the Marine Park manager, Chief Ranger and Rangers who are all employed by STINAPA Bonaire. The mission of the BNMP is 'protect and manage the island's natural, cultural and historical resources, while allowing ecologically sustainable use, for the benefit of future generations'. It hosts approximately 65,000 visitors a year of which 40,000 are SCUBA divers. One of the primary challenges of managing the BNMP is dealing with the various stakeholders who use the park.

Scenic assessment of the 18 coastal locations that comprises the island indicates that 61% are included in Classes I and II (10 and 1 site, respectively). Three (17%) belonged to Class III, and four of the sites (22%) corresponded to Classes IV (3) and V (1). Some examples of scenic analysis (Chap. 4) shown in Fig. 6.19 are taken from the 100 Steps Beach locality, Donkey Beach area and the Plaza Resort.

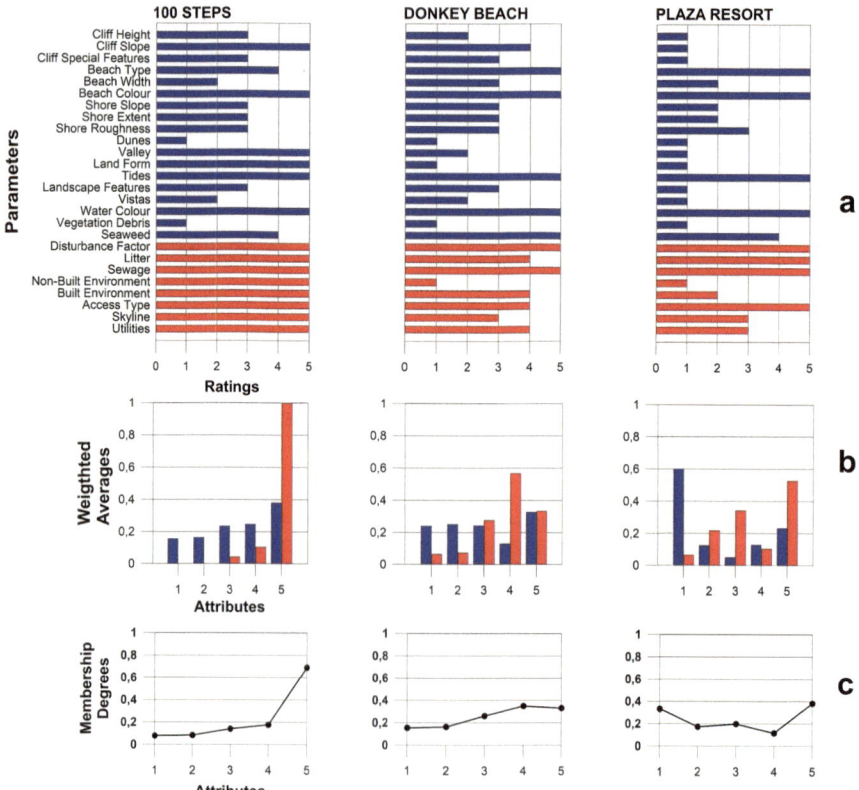

Fig. 6.19 (**a**) Scenic evaluation histograms; (**b**) Weighted averages and (**c**) Membership degree curves for 100 Steps (Class I, D: 0.96), Donkey Beach (Class III, D: 0.45) and Plaza Resort (Class V, D: -0.04)

The 100 Steps obtained its name because of the limestone stairs that people have to descend to arrive at the beach (Fig. 6.19). It is located inside the BNMP, and its attractiveness is linked to clear water conditions and bleached pebbles having a biogenic origin (mainly coral). Highlights are the natural vegetation cover and limestone cliffs (Fig. 6.20a). Despite the easy access, visitor influx is low bringing perfect scores to all human related parameters. This location is ideal for diving and snorkelling activities, because a wide variety of fish and turtles are often found in the shallow, turquoise waters.

Donkey Beach presented high scores in physical parameters due to a very clear turquoise water, white sediments of biogenic origin and a great developement of mature trees (Figs. 6.19 and 6.20b). However, the beautiful scenic setting decreases because it is located in front of the airport and adjacent to a bunkering port. It is popular amongst locals, who often bring their whole family for a weekend BBQ. The parking lot is visible from the beach, and food trucks are often present.

Plaza Resort is a private area essentially frequented by hotel guests and cruise visitors. Severe erosion related to extreme wave events and hurricanes has significantly reduced beach width in this area (Fig. 6.20c), as well as in some parts of the island (Fig. 6.20d). Despite very clear turquoise water and a superb sand beach colour, the remaining natural parameters are quite poor (Figs. 6.19 and 6.20c). Noise disturbance is not important due a strict control by hotel administration and both vegetation debris and beach litter is collected twice a day. Skyline and utilities show poor scores because of the presence of hard coastal defenses emplaced by various hotel administrations to protect beach areas and their related facilities from direct wave attack and coastal erosion (Figs. 6.19a and 6.20c).

Fig. 6.20 Examples of sites along Bonaire coastline: (**a**) 100 Steps (Class I); (**b**) Donkey Beach (Class III); (**c**) Plaza Resort (Class V) and (**d**) example of severe coastal erosion at this site

6.3.2 Colombia: Caribbean Coast

This is a developing area divided into eight Departments including 28 Municipalities, with 4,049,867 inhabitants mainly concentrated in four commercial and tourist cities: Barranquilla, Cartagena de Indias, Santa Marta and Riohacha. Due to better security conditions and the recent Peace agreement, this area has become a popular destination for many national and international tourists attracted by tropical rural, remote and urban sites beach-related activities and local entertainment. At present, tourism represents one of the most important economic activities and its development capacity is experiencing sound growth.

Scenic assessment of 150 locations showed that 31% of the investigated coastal areas were included in Class I (25 sites) and Class II (22), 16% belonged to Class III (24) and 47% of the sites corresponded to Classes IV (34) and V (45). Classes I and II sites are areas with spectacular water and sand colour, luxurious vegetation, etc. They are usually located in Natural Parks and are especially abundant along sectors where the mountainous relief gives rise to striking landscape characteristics enhanced by the presence of unique geological features. Sites categorized within Classes III, IV and V are recorded in locations where litter and sewage evidence are frequently observed (Williams et al. 2016; Rangel-Buitrago et al. 2017 and 2018). At many sites, degradation was enhanced by erosion processes, which have been counteracted by hard engineering structure construction that further reduced scenic value (Rangel-Buitrago et al. 2015, 2017 and 2018).

Coastal management competences belong to three governmental institutions, i.e. the Colombian Oceanic Commission (CCO), the General Maritime Directorate (DIMAR) and the Ministry of Environment and Sustainable Development (MADS) and two councils, e.g. the National Environmental Council (CNA) and the National Council of Economic and Social Politic (CONPES). Despite the many Institutions involved, it is not clear which Institution is directly responsible for addressing coastal scenery; similarly for the particular interferences and functions associated with this valuable resource. Advances regarding coastal management are very weak and considering the high potential observed along the Caribbean coastline, current results remain unclear (Avella et al. 2009).

Some examples of scenic analysis (Chap. 4) are shown in Fig. 6.21, namely Arrecifes, Playa Blanca and Galerazamba sites.

Arrecifes is a remote area located in Tayrona Park, sited on the northern Caribbean coast. Its attractiveness is linked to clear water conditions, mature natural vegetation cover and golden coloured sand derived from weathering of granite rocks located in the small drainage basin that feed this locality (Fig. 6.22a). The 'skyline landform' parameter is mountainous due to the Sierra Nevada de Santa Marta, an isolated mountain considered as the world's highest coastal range positioned just 42 km from the coastline. Natural Park status (protected area) implies that human impacts are low – almost zero – generating excellent scores for all human parameters (Fig. 6.21).

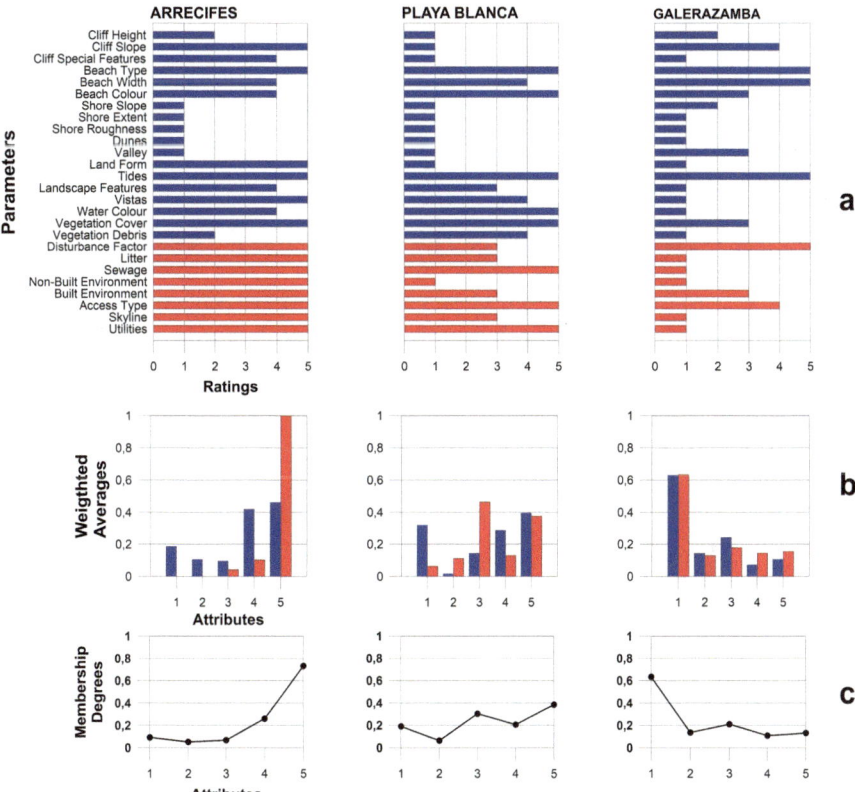

Fig. 6.21 (**a**) Scenic evaluation histograms; (**b**) Weighted averages and (**c**) Membership degree curves for Arrecifes (Class I, D: 1.20), Playa Blanca (Class III, D: 0.63) and Galerazamba (Class V, D: -0.65)

Playa Blanca is located adjcent to Corales del Rosario and San Bernardo Natural Parks (Fig. 6.22b). This area presents high scores thanks to very clear turquoise water, white sediments of biogenic origin and a great developement of mature trees (Fig. 6.21). However, the attractive scenic setting is severely affected by activities related to mass tourism that generates noise, litter and construction of bungalows to cope with the tourism demand (Figs. 6.21a and 6.22b).

Galerazamba is a village area located on the central Caribean coast. The natural parameters are quite weak and especially remarkable are the muddy brown water-colour and high litter amounts that reach values of 7.48 items m^{-1}, both associated to Magdalena river fluxes, the largest river system in Colombia (Figs. 6.21 and 6.22c, d). Sewage, skyline and utilities show poor scores due to alteration as a result of seawall construction by local house owners (Figs. 6.21a and 6.22c, d).

Fig. 6.22 Examples of sites along the Caribbean coastline of Colombia: (**a**) Arrecifes (Class I) located at Tayrona National Park; (**b**) Playa Blanca (Class III) and (**c–d**) Galerazamba (Class V)

6.3.3   *Cuba*

The Cuban archipelago consists of Cuba Island, the Island of Juventud and Sabana-Camagüey, Canarreos and Jardines de la Reina Archipelagos. Modern tourism development in Cuba started at the end of the 1980s to counteract the economic crisis due to the Soviet Union dissolution and the USA embargo (Cerviño and Cubillo 2005).

International coastal tourism is concentrated at cays and at Varadero where national tourism is acquiring greater importance. Urban and tourism developments are firmly regulated by laws/directives. Cuban Constitution (article n. 27) promotes the rational use of natural resources and Environmental Laws n. 33 (1981) and n. 81 (1997) regulates environmental protection and preservation. As regards coastal developments, specific Law Directive n. 2012 was enforced in 2000 under the supervision of the Cuban Ministry of Science, Technology and Environment (UNEP/GPA 2003).

Construction is forbidden within 40–80 m from the shoreline and on sand dunes. Since 1987, at Varadero and other selected sites, coastal occupation took place according to a sound Coastal Management Programme that allowed building construction 40 m behind dune ridges. Beach nourishment/dune reconstruction was preferred to hard solutions to avoid environmental and scenic degradation processes (Anfuso et al. 2014), although a darkening in sand colour has occurred (Pranzini et al. 2016).

One hundred and one locations were analysed. Class I scenic sites were usually located close to mountainous chains in the north-western and south-eastern coast of

Cuba Island (Anfuso et al. 2017). Cays present a stunning water colour, white sand and luxuriant vegetation but no cliffs, rock shores and mountainous skyline and, for these reasons, are not Class I but II sites. Examples of scenic analysis (Chap. 4) shown in Fig. 6.23 with scenic evaluation histograms (a), weighted averages (b) and membership degree values (c) together with D values leading to class decisions, are represented by Playa Colorada, Playa Cayo Ensenacho and Mayabeque.

Playa Colorada lies in a striking, remote area located in southern-east Cuba whose attractiveness is linked to the presence of cliffs and a rocky shore with vibrant yellow-red sand colouring due to iron oxide content resulting from limestone weathering in the small watershed feeding this area (Fig. 6.24a). Skyline landform is undulating due to the Sierra Maestra mountainous chain. Natural vegetation, composed by mature trees, rate highly (Fig. 6.24a) and human impact is small – all human parameters showing very good scores (Fig. 6.24a) leading to a Class I rating.

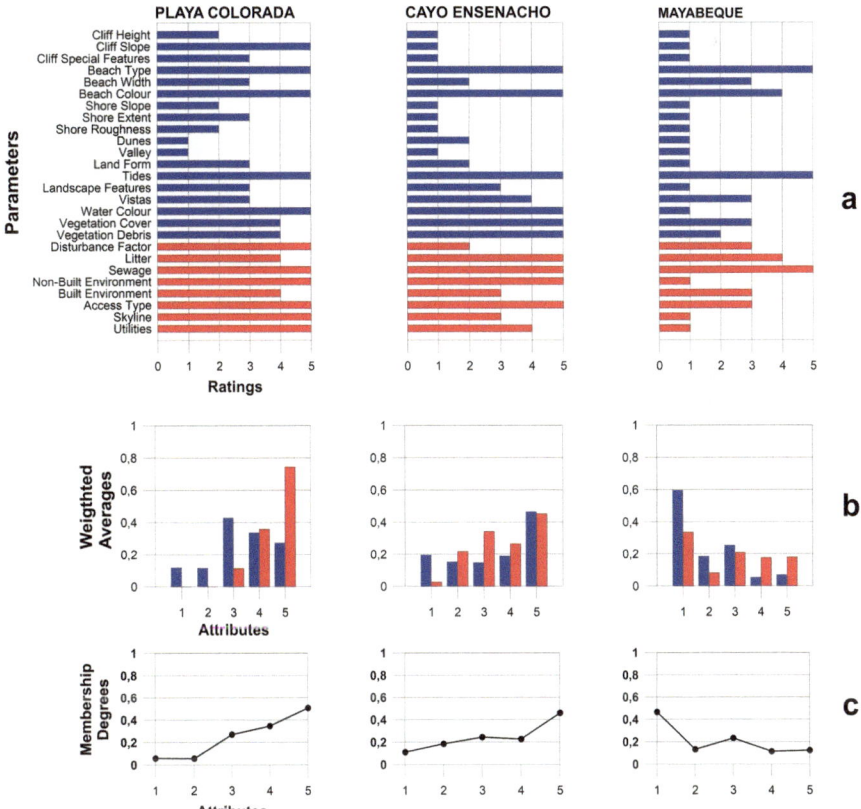

Fig. 6.23 (**a**) Scenic evaluation histograms (**b**) Weighted averages and (**c**) Membership degree curves for Playa Colorada (Class I, D: 0.92), Playa Cayo Ensenacho (Class III, D: 0.43) and Mayabeque (Class V, D: -0.48)

Fig. 6.24 Examples of locations along the Cuba coastline: (**a**) Playa Colorada (Class I); (**b**) Playa Cayo Ensenacho in a resort area (Class III) and (**c**) the same beach in a natural area and (**d**) Mayabeque (Class V)

Playa Cayo Ensenacho is backed by an international resort located in the cay in Northern-central Cuba. Beach attractiveness is linked to natural vegetation cover, dunes, the turquoise water colour and white bright sand (Figs. 6.23 and 6.24b, c). Vegetation debris is virtually absent because of continuous manual cleaning operations. With respect to the human parameters, noise disturbance is a problem due to loud music (for tourist entertainment) on the beach (Fig. 6.23) leading to its Class III rating. Built environment and skyline parameters obtain an intermediate score because bungalows are emplaced behind dune ridges and masked by luxuriant vegetation (natural and planted, Fig. 6.24b), which also constitutes (along with dunes) an effective buffer zone between beach and backing structures/activities. Litter is virtually absent because of the afore-mentioned cleaning operations. Massive and/ or permanent presence of beach umbrellas, sun chairs, negatively affect scenic beauty – this is evident by comparing the analysed sector with the adjacent natural one (Figs. 6.24b, c).

Located on the flat south-western coast of Cuba, is Mayabeque, a village area essentially frequented by local beachgoers. Strong erosion, because of several hurricanes, has greatly reduced beach width (Fig. 6.24d). Natural parameters are generaly poor, but especially remarkable is the unattractive dark water colour due to staining by organic acids from streams crossing backing mangrove swamps (Fig. 6.24d). Noise disturbance is not important because of low visitor numbers and beach litter is almost daily collected (Fig. 6.23). Skyline and utilities show poor scores because of the presence of hard, chaotic defence structures emplaced – without an appropriate management plan – by owners to protect their houses and/or restaurants (Figs. 6.23 and 6.24d).

6.3.4 Morocco: Mediterranean Coast

The Morocco coastline (3411 km) consists of cliffs (c. 2131 km) covering c. 63% of the total coastline length), beaches (958 km, c. 28%), coastal lagoons (68 km, c. 2%) and estuaries and deltas (254 km, c. 7%). Along the 512 km Mediterranean coast, cliffs occupy 80% of its length, with beaches the remaining 20%, these being essentially located on low coastal plains. Some 45 sites are important tourist areas, as during summer they receive on average 1500 beachgoers per day and water quality is permanently monitored. Fifty-five percent of the population (18.3 million) live in coastal areas, especially along the Mediterranean coast. This area has the greatest tourism pressure and where the bulk of the population is employed in the tourism industry, it represents the economic engine of Morocco. Dense coastal population exerts an environmental cost, e.g. coastal erosion, beach sand mining for the construction industry, sewage and waste material disposal. Tourism is threatened by coastal erosion, which ranges from 1.1 m/year (for the eastern sector, Boumeaza et al. 2010), 1.7 m/year (central, Nachite 2009) to 1.8 m/year (western sector, Anfuso et al. 2010).

The Department of the Environment is responsible for the national policy for environmental management, and many different agencies implement various policies often independent of one another – not a very efficient system. A Coastal Act (n° 81–12) was approved in July 2015 but still has not been applied. Deficiency of social and organizational system can be seen as the main challenges to be overcome for adequate management of the coast, a challenge that emerges in all Integrated Spatial Planning programmes/projects. This deficiency appears mainly in: poorly adapted instruments and tools; a weak coordination between coastal management administrations and lack of adequate training of coastal managers.

Some examples of scenic analysis (Chapter 4; Williams and Khatabbi 2015) are taken from Kamkoum el Baz (Ras Kebdana), Al-Kallat and M'diq (Figs. 6.25 and 6.26).

Kamkoum El Baz, or Ras Kebdana, is a remote site and obtained the highest scenic value for 24 investigated coastal locations. The 8 km coastal frontage is fronted by a spectacular cliff backed by migrating foredunes and well vegetated dunes (Figs. 6.25 and 6.26a, b). Cleaning litter that has been produced by some 4,000 visitors per day is done manually and no beach bins were evident. Anthropogenic parameters all scored an attribute 5 apart from litter. Some 275 litter

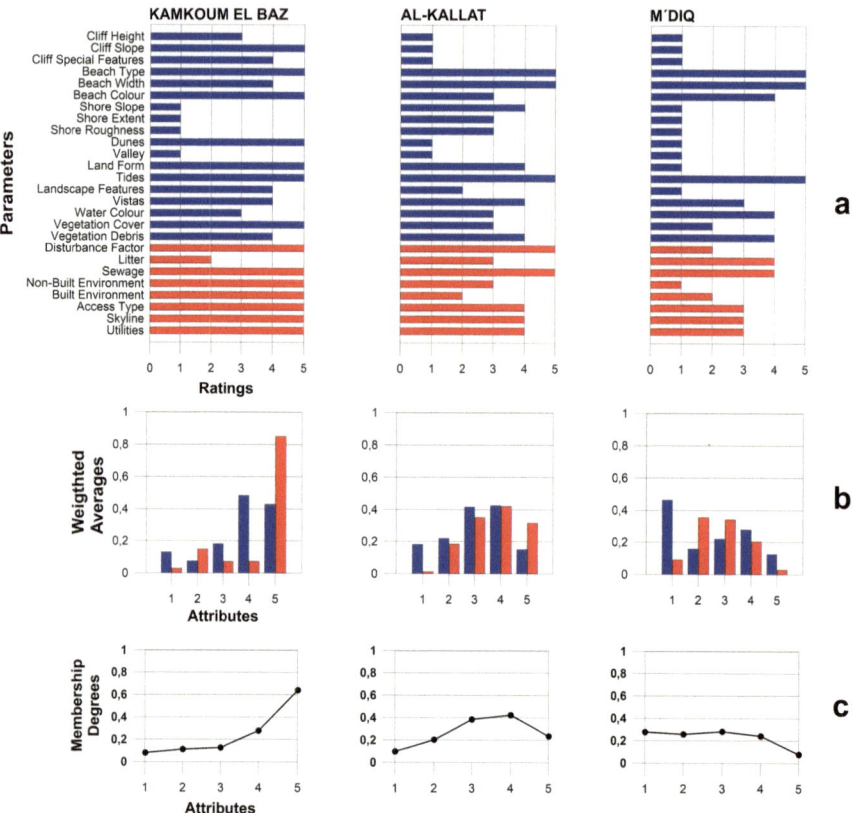

Fig. 6.25 (**a**) Scenic evaluation histograms; (**b**) Weighted averages and (**c**) Membership degree curves for Kamkoum el Baz (Class I, D: 0.92), Al Kallat (Class III, D: 0.40) and M'diq (Class V, D: -0.23)

items (mainly plastic)/100 m of beach were observed, a ubiquitous problem for the whole of the Moroccan coast.

Al-Kallat is classed as a resort, but tourism is mainly by nationals and construction has been ongoing for years with a new port geared to be built south of the town. Some 433 general litter items/100 m and four gross litter items were counted and this, in conjunction with the built/non built environment and rise of construction, played a large part in its Class III grade. The actual beach area– as in most of the locations studied, scored highly on attributes (Fig. 6.25). A tourist attraction is a small coastal freighter that ran aground 1.5 km north of Al-Kallat (Fig. 6.26c).

The urban area of M'diq, on the westernmost Mediterranean coast, has greatly expanded its tourism potential due to its vicinity to Tetuan, the Gibraltar area and Spain. All the coast between Ceuta and the Cabo Negro promontory has experienced massive development in recent decades with construction of a motorway, condominiums, summer houses, hotels and three ports, i.e. Marina Smir (in 1986),

Fig. 6.26 Examples of beaches along the Mediterranean coast: (**a–b**) Kamkoum el Baz, Class I; (**c**) Al-Kallat, Class III and (**d**) M'diq (Class V)

Marina Kabila (1991) and M'diq fishing port (between 1961 and 1966). The natural parameters of the latter gave good scores for beach characteristics and water colour (Fig. 6.26d). Human structures are emplaced on dune ridges and greatly affect vistas and skyline landforms. Noise disturbance is a problem – loud music, sports and in general the sheer number of visitors.

6.3.5   Spain: Andalusia, Basque Country and Catalonia

The Spanish coastline exhibits an almost rectilinear form. The tectonic character of the coast gives rise to mountain chains often parallel to the coastline that usually presents an abrupt topography. Fifty percent of the entire coastline (which is 7870 km) is composed of cliffs especially abundant on the Atlantic (Galicia, Asturias, Cantabria, and the Basque Country) and Mediterranean (Catalonia and eastern Andalusia) coasts. At numerous places cliffs give rise to prominent headlands confining pocket beaches of great scenic value. Sand beaches and low lying areas respectively constitute 25% and 17% of the Spanish coast with built areas occupying 8%. Today more than 35% of the Spanish population lives within a coastal belt of a 50 km width. Erosion processes were counteracted during the 1960s

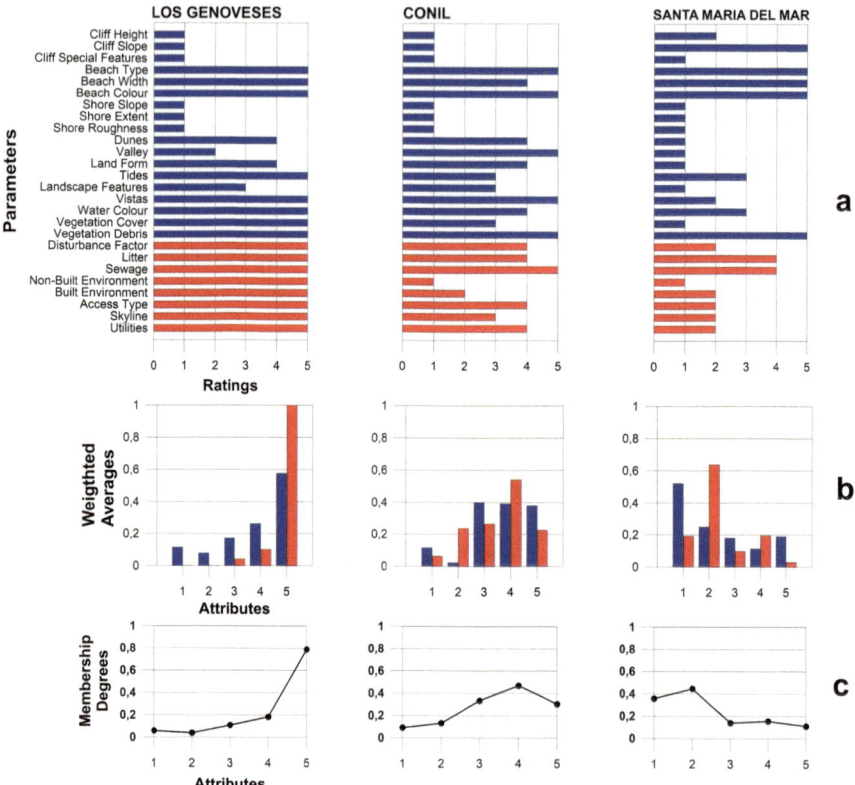

Fig. 6.27 (**a**) Scenic evaluation histograms; (**b**) Weighted averages and (**c**) Membership degree curves for Los Genoveses (Class I, D: 1.26), Conil de la Frontera (Class III, D: 0.64) and Santa María del Mar (Class V, D: -0.70)

and 1970s by groin and breakwater constructions, whilst artificial nourishment projects started in the 1980s (Manno et al. 2016). Such increasing coastal urbanization has deeply affected coastal scenery, especially along the Mediterranean areas, i.e. in Catalonia and Andalusia (Williams et al. 2012), which are among the most densely populated areas.

In this study, 94 sites were classified along the coastlines of the Basque Country, Catalonia and Andalusia. Basque Country sites had good scores due to good natural parameters and low human impacts, in comparison to the Catalonia coast and the Andalusia coastline presented a great variety of cases (Fig. 6.27). Despite the presence of very urbanised areas on the Mediterranean side, e.g. Costa del Sol (Malaga area), it has c. 32% of the coastline under protection. Protected areas, essentially located on the Atlantic side, are under local, regional and national protection features and RAMSAR and Natura 2000 networks. Examples of scenic analysis (Chap. 4) shown in Fig. 6.27 are taken from Los Genoveses, Conil de la Frontera and Santa María del Mar.

Los Genoveces is located in a remote area in eastern Andalusia within the Cabo de Gata-Nijar Natural Park, a spectacular mountainous setting of Tertiary volcanic rocks. Such formations give rise to two headlands protruding into the Mediterranean Sea (Figs. 6.28a, b), enclosing the site and forming a 1.4 km long pocket sand beach backed by wide and well developed dune ridges (giving good scores at points 4–6 and 10, Fig. 6.27). Los Genoveses is emplaced in a large, dry valley and shows a highly undulating landform, vistas open on four sides and several special landscape features (Fig. 6.27). The water colour is turquoise and clarity is excellent. The vegetation cover is well developed and vegetation debris is virtually absent. All human parameters score very highly (Fig. 6.27).

On the Atlantic coast of Andalusia, the attractive setting of Conil de la Frontera is affected by human impacts. The wide beach is composed of golden coloured quartz sand has well developed dune ridges and a river mouth (Figs. 6.27 and 6.28c). Intermediate scores were recorded for other natural parameters, such as, valley, land form, landscape features and vistas (Fig. 6.27). Human parameters gave average values, e.g. litter, as daily manual/mechanical cleaning operations occur, which keep the litter content low despite the intense visitor influx, but low scores were recorded at several points due to construction activity (Figs. 6.27 and 6.28c).

Santa María del Mar, on the Atlantic coast of Andalusia, is an urban area that has undergone beach nourishment several times in recent decades. With respect to natural parameters, good scores were obtained for cliff, beach width and sand colour, but scores for the other natural parameters were poor (Fig. 6.27). Despite litter and sewage evidence having good scores because of daily cleaning operations

Fig. 6.28 Examples of sites along the Andalusia coastline: (**a–b**) Los Genoveses (Class I); (**c**) aerial view of Conil de la Frontera (Class III) and (**d**) aerial view of Santa María del Mar (Class V)

during the summer period, other human parameters gave low scores; the disturbance factor is high due to the great influx of visitors (carrying capacity is often exceeded), the natural environment has been deeply transformed and several utilities can be observed (Fig. 6.28d, e.g. groins that limit the beach and a revetment that covers most of the backing cliff).

6.3.6   Turkey: Mediterranean Coast

The turquoise and sparkling waters of the Turkish Mediterranean and Aegean coasts have become the centre of a rapidly growing tourism sector following the Tourism Incentives Law enacted in 1983 by the Ministry of Tourism and Culture. The Environmental Law (1983), Amendments (1986 and 1998) and Decree for Specially Protected Areas (SPA 1982) by the Ministry of Environment (renamed Ministry of Environment and Forestry in 2003) was enacted to cover environmental issues relating to the adverse impacts of tourism development. It dealt with issues, such as, pollution, noise, water quality, solid waste management and environmental impact assessment providing rules and regulations for environmental management. Reports on the Tourism Strategy for Turkey-2023 (Ministry of Culture and Tourism, Ankara, 2007) aim to improve the tourism sector, not on an individual land basis but regionally, that targets a wiser use of the natural, cultural, historical and geographical characteristics of the Turkish coast. The strategic plan mainly emphasizes conservation, protection and sustainability together with coastal eco-tourism based on tourism research with a development strategy in cooperation with involving participation of the public and private sector, NGOs and universities.

The Coastal Scenic Evaluation System (Chap. 4) is exemplified by scenic analysis of Çıralı Mid Town, Göksu Hurma and Antalya Dedeman Hotel all located on the Mediterranean coast. Their membership degree values, weighted averages and D values, leading to their class divisions, are represented in Fig. 6.29a.

Çıralı Mid Town coastal site is located 70 km west of Antalya. The Taurus Mountains surround the Gulf in the north where a natural fire known as Yanartaş (Burning Rock-Chimera) burns continuously. Çıralı is a place of historical treasures from Lycian, Hellenistic and Roman times and was rated highly due to outstanding features in both natural and human parameters (Fig. 6.30a). The final rating (D: 1.31) was strongly influenced by the striking historical features plus a dense vegetation cover and an optimal beach type that even serves as a breeding place for the turtle *Caretta caretta*. The top five rated human parameters obtained from the scenic analysis were the absence of sewage, water colour/clarity, the absence of noise, quality of the built environment, all scoring an attribute value of 5, the lowest attribute score being 4 for litter (Ergin 2009).

The Göksu Hurma site is located at the apex of a river delta near Mersin. This site was highly rated on beach parameters, in spite of having no rocky shore and cliffs. No urban development occurred in the area and it scored highly on human parameters. However, the absence of special landscape features and considerable amounts of litter resulting in a Class III rating (Fig. 6.30b).

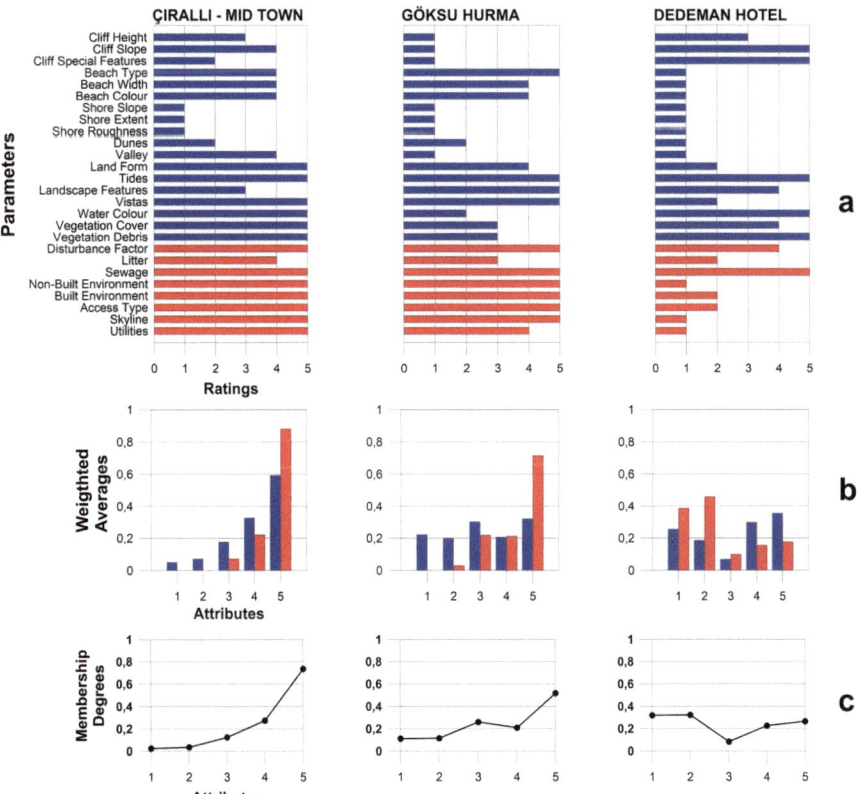

Fig. 6.29 (**a**) Scenic evaluation histograms; (**b**) Weighted averages and (**c**) Membership degree curves for Çıralı Mid section, Kemer (Class I, D: 1.31), Göksu Hurma, Mersin (Class III, D: 0.61) and Dedeman Hotel, Antalya (Class V, D: -0.21)

The Dedeman Hotel site is located within the town of Antalya in a superb area having cliffs and clear water. Its attribute rating peaks at the cliff parameters and no rocky shore and beach were evident. The site typically had features such as unattractive urbanization, ugly coastal structures, intensive tourism (hotels) development with many ugly utilities. The skyline surrounding the site is unattractive, downgrading the natural environment (Fig. 6.30c).

6.3.7   Wales, United Kingdom

In 1973/74, three pilot schemes (Sussex, Dorset and Glamorgan) for Heritage Coasts (HC) were initiated in England and Wales. The HC aim was to protect coastlines of unique scenic beauty, and environmental value i.e. quality was emphasized (Williams 1987, 1992; Phillips et al. 2010). If natural beauty and recreation were ever to be in conflict, then the former should prevail, but only where efforts to

Fig. 6.30 Examples of sites along the Turkish coastline: (**a**) Çıralı-Mid section, Kemer (Class I); (**b**) Göksu- Hurma Mersin (Class III) and (**c**) Dedeman Hotel, Antalya (Class V)

resolve the conflict by good planning and management have failed. Currently 45 HCs exist and the management philosophy prevails over 1500 km of the England and Wales coastline. Seven HCs covering over 35% of the coastline, cover 14 distinct coastal stretches and 495 km are in Wales (population 3.1 million; area 2.1 m ha) and the coast, which extends for 1300 km from Chepstow to Queensferry, had a full coastal footpath inaugurated in 2012. The status is non-statutory, which leaves them vulnerable in times of financial stringency but local planning authorities must take cognizance of these areas before taking any decisive action on matters affecting the coastal area.

Scenic beauty including not only the physical aspects of the landscape, i.e. geology, geomorphology, flora, fauna, etc. but also cultural heritage features of historical and archaeological interest, is the primary focus. A fundamental tenet of the conservation philosophy is local involvement. It is low-key management approach epitomized by discrete signage of the area with the aim of letting people find out about the area themselves. School visits, 'adopt a beach' campaigns, ecotourism, booklet production, pragmatic academic research, amonts others, are all facets of the management scheme.

Examples of scenic survey results (Chap. 4) shown in Fig. 6.31 (a), scenic evaluation histograms, (b) membership degree and (c) weighted averages are taken from the Ceredigion Heritage Coast (Poppit) in West Wales and the Glamorgan Heritage Coast (Southerndown and Gileston just outside the Heritage coast) in South Wales.

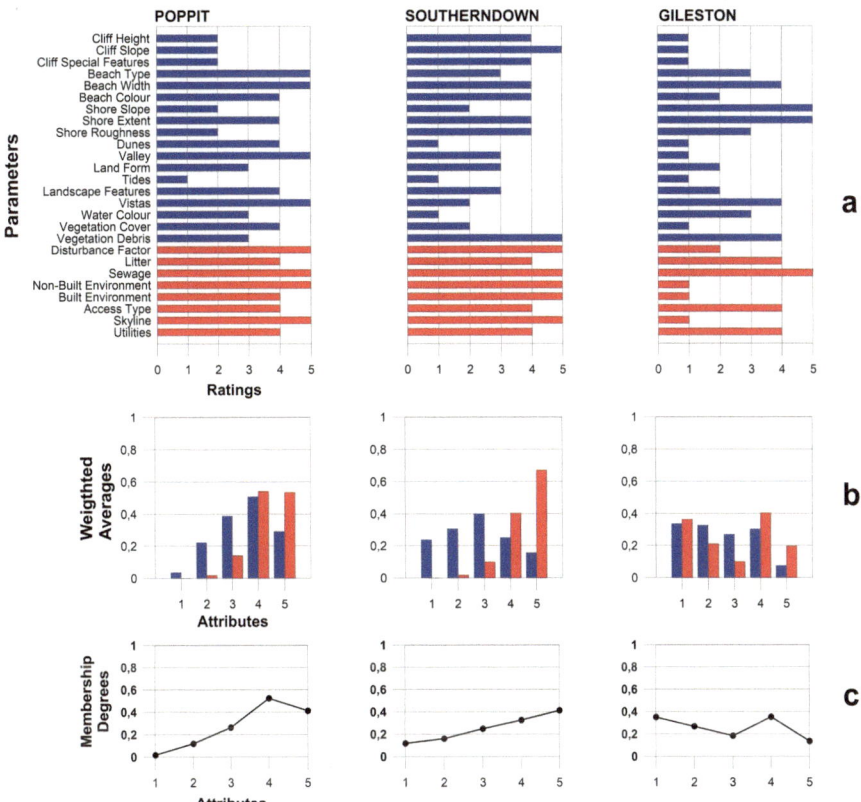

Fig. 6.31 (a) Scenic evaluation histograms; (b) Weighted averages and (c) Membership degree curves for Poppit (Class I, D: 0.89), Southerndown (Class III, D: 0.54) and Gileston (Class V, D: -0.08)

Fig. 6.32 Examples of locations along the Welsh coastline. (**a**) Poppit (Class I); (**b–c**) Water jetting site at Southerndown (Class III) and (**d**) Gileston (Class V)

Southerdown is classed as a III site even though it lies in a Heritage Coast. It is unrealistic to have every site in an Heritage area or National Park as a Class I site and this is the case with Southerndown.

Poppit sited in the only coastal National Park in England and Wales, is located on the River Teifi near Cardigan town scored highly on human parameters, the lowest attribute score being 4. Low cliffs were seen, and vistas were excellent, e.g. to Gwbert and Cardigan Island, as was the long sand beach itself and adjacent dune complex (Fig. 6.32a). Management could look into the issue of removing strandline debris to increase the D value and to improve Class I characterstics.

Southerndown has superb Carboniferous/Triassic cliff scenery and a large Iron Age fort site, whose fortification ridges can be seen, sited on the eastern cliff top (Fig. 6.32b). The scar left by a failed water jetting experiment a bad management decision by the local council to counteract blocks falling to the beach with the risk of hitting beach sun bathers, can be clearly seen on the cliff face (Fig. 6.32c). The jetted area now has retreated at three times (~30 cm/year) the rate of the adjacent cliff. The beach is a vast expanse of dark tan sand backed by bleached cobbles, but strandline marine debris again causes a problem. Human parameters are excellent, the lowest attribute value being 4 and with a little effort this site is close to becoming a Class II site although presently classified as a Class III.

Gileston, is a small village and has a very extensive cobble/pebble beach – it is a 'sink' for longshore eastward drifting sediment and the main factor contributing to its Class V value is the power station and large sea wall (Fig. 6.32d). The water colour is also poor due to silt in the adjacent waters of the Bristol Channel. Examination of the histogram shows many low scores in the appropriate attribute columns (Fig. 6.32c).

6.4 Conclusions

Newton with his prism and silent face… is voyaging through strange seas of thoughts, alone. (Wordsworth 1896, 61)

In the first section is presented the distribution of 952 sites around the world characterised from a scenic viewpoint using the method presented in Chaps. 4 and 5. The second section provides examples from different countries of the five investigated classes, from Class I (very attractive sites) to Class V (unattractive urban areas). The third section presents large areas and entire countries analysed in depth. Such areas, i.e. the Caribbean coast of Colombia, the Mediterranean coast of Morocco, Andalusia region in Spain, Mediterranean Turkey and countries, i.e. Bonaire, Cuba and Wales, have been briefly described giving information on their physical and scenic characteristics as well as on environmental policies and management issues in force in each case.

Class I sites need outstanding scenic qualities quite often linked to several physical parameters. Cliff, rock shore, valley, skyline landform and (partially) coastal landscape features and vistas are frequently linked to the presence of a high rocky coast and (often) an undulating or mountainous landscape.

Erosive marine and aerial processes affect a high coastal sector favouring the formation of an eroding cliff/bluff and/or a rock platform cut into the cliff base. Soft rocks give rise to rapidly retreating cliffs or bluffs and no rocky shore formation. A resistant and hard horizontal stratified rock is prone to form a vertical cliff fronted by a flat rock platform. Geological characteristics affect/control cliff retreat favouring the formation of coastal special features, such as, rock ridges, irregular headlands, stacks (Fig. 6.33a), arches (Fig. 6.33b), shore platform (Fig. 6.33c), notches (Fig. 6.33d), etc. which give very high scores to any site.

Natural vegetation is an important component of scenery and the presence of luxuriant vegetation on the backbeach makes a site very attractive. The above mentioned parameters invariably obtained good scores at Class I and II sites, which are not affected by human impact because they are usually located in (or close to) natural protected areas. It is interesting to highlight some places, e.g. Varadero and several beaches in the Cuban cays, where an appropriate dune management has provided an effective buffer zone between the beach and the backing human structures, activities, favouring the presence of Class II sites in resort and even urban areas. Similarly, white sediment influences water colour and gives it a very attrac-

Fig. 6.33 Examples of coastal special features, which give very high scores to the site. (**a**) stacks (Portugal), (**b**) arches (Iceland, photo by Andy Jones), (**c**) shore platforms (Colombia), (**d**) notches (Bonaire)

tive clear and turquoise hue, but such a valuable characteristic, very much appreciated by beachgoers, does not necessarily give rise to a Class I beach, as was observed in Cuba at many cays usually rated as Class II.

Classes III and IV represent the most common situations and almost cover 50% of all ranked sites. They are the result of low quality physical parameters and the increasing influence of anthropogenic impacts, such as, noise disturbance, presence of litter, sewage evidence and low sensitive human constructions.

As an example, noise is linked to traffic and human activities in the backbeach area and loud music is often common in urban beaches because it is linked to local beach users and the presence of bars and restaurants. At resort areas, it is used as entertainment for tourists but can also be linked to nautical sports, such as, jetskis, 'bananas', etc.

Litter and evidence of sewage discharge are linked to different items abandoned on the beach by visitors or transported onto the shore by rivers, waves and currents. Sound management and educational initiatives can control beach litter, e.g. by placing an adequate number of bins that are emptied regularly and improvement of appropriate manual and mechanical cleaning operations; noise disturbance can also be reduced by good management.

Class V sites are usually very common and include urban beaches very negatively affected by human impact because of the total absence of buffer zones, the presence of high traffic, high buildings, poor facilities.

At many places, not only at urban but also village and rural areas, low scenic scores are linked to coastal erosion processes that favour beach reduction, sediment coarsening, dune erosion, vegetation degradation, bad building design and, unfortunately, the chaotic emplacement of defence structures that further low scenic value. In such cases, beach nourishment should be preferred to hard defence solution to improve scenery but also beach carrying capacity and beachgoers safety.

Appendix 1: Worldwide Coastal Locations Surveyed

No	Coastal site name	Country	Continent	D value	Class
1	Salisbury Plain. South Georgia	Antarctica	Antarctic	1.25	I
2	Carcass island, Falklands	Antarctica	Antarctic	0.75	II
3	Ushuaia	Antarctica	Antarctic	0.30	IV
4	Ship cove, West Falklands	Antarctica	Antarctic	0.39	IV
5	Gritviken	Antarctica	Antarctic	0.87	I
6	Shackleton	Antarctica	Antarctic	0.99	I
7	Pleasant cove	Antarctica	Antarctic	0.43	III
8	Gold Harbour	Antarctica	Antarctic	1.00	I
9	Arctowski	Antarctica	Antarctic	0.73	II
10	Deception island	Antarctica	Antarctic	0.75	II
11	Austenmeer Beach	Australia	Oceania	0.62	III
12	Dee Why	Australia	Oceania	0.86	I
13	Manley	Australia	Oceania	−0.13	V
14	Long Reef	Australia	Oceania	1.39	I
15	Bondi	Australia	Oceania	0.27	IV
16	Sorobon Beach	Bonaire	America	0.42	III
17	Atlantis Beach	Bonaire	America	0.94	I
18	Pink Beach	Bonaire	America	1.02	I
19	Donkey Beach	Bonaire	America	0.45	III
20	Palu di mangel Beach	Bonaire	America	0.19	IV
21	Bonaire plaza resort south	Bonaire	America	0.20	IV
22	Bonaire plaza resort beach south	Bonaire	America	−0.04	V
23	Coco Beach	Bonaire	America	0.22	IV
24	Andrea Beach	Bonaire	America	0.52	III
25	1000 Steps beach	Bonaire	America	0.96	I
26	Tolo Beach	Bonaire	America	0.99	I
27	Playa Frans	Bonaire	America	0.95	I
28	Boka slagbaai	Bonaire	America	0.75	II

(continued)

No	Coastal site name	Country	Continent	D value	Class
29	Wayaka beach	Bonaire	America	1.35	I
30	Playa Funchi	Bonaire	America	1.31	I
31	Playa Benge	Bonaire	America	1.29	I
32	Playa Kokolishi	Bonaire	America	1.32	I
33	Playa Chikitu	Bonaire	America	1.40	I
34	Parque da Guarita	Brasil	America	0.52	III
35	Praia da Cal	Brasil	America	0.05	IV
36	Praia dos Molhes	Brasil	America	0.04	IV
37	Parque da Itapeva Sul	Brasil	America	−0.02	V
38	Balneário Itapeva Sul	Brasil	America	−0.03	V
39	Prainha	Brasil	America	−0.06	V
40	Parque da Itapeva Centro	Brasil	America	−0.14	V
41	Balneários Centro-sul	Brasil	America	−0.22	V
42	Balneários Extremo Sul	Brasil	America	−0.25	V
43	Praia Grande	Brasil	America	−0.36	V
44	Parque da Itapeva Norte	Brasil	America	−0.42	V
45	Praia Grande Sul	Brasil	America	−0.46	V
46	Arroio Corrente	Brasil	America	−0.31	V
47	Arroio Corrente Sul	Brasil	America	0.26	IV
48	Arroio do Silva Centro	Brasil	America	−0.59	V
49	Arroio do Silva Norte	Brasil	America	−0.03	V
50	Arroio do Silva Sul	Brasil	America	0.18	IV
51	Balneário Gaivota Centro	Brasil	America	−0.34	V
52	Balneário Gaivota Deserta	Brasil	America	0.20	IV
53	Balneário Gaivota Norte	Brasil	America	−0.17	V
54	Balneário Gaivota Sul	Brasil	America	0.20	IV
55	Balneário Morro dos Conventos	Brasil	America	0.22	IV
56	Balneário Rincão Centro	Brasil	America	−0.46	V
57	Balneário Rincão Sul	Brasil	America	0.19	IV
58	Balneários Centro Jaguaruna	Brasil	America	−0.27	V
59	Barra da Ferrugem	Brasil	America	0.24	IV
60	Barra Velha Araranguá	Brasil	America	−0.22	V
61	Barra Velha Rincão	Brasil	America	−0.22	V
62	Bela Torres	Brasil	America	−0.29	V
63	Camacho	Brasil	America	−0.15	V
64	Camacho Sul	Brasil	America	0.35	IV
65	Cardoso Norte	Brasil	America	0.12	IV
66	Cardoso Sul	Brasil	America	0.42	III
67	Cigana	Brasil	America	0.44	III
68	Dunas do Sul	Brasil	America	0.06	IV
69	Figueirinha	Brasil	America	0.35	IV
70	Galheta Aberta	Brasil	America	0.45	III
71	Galheta Norte	Brasil	America	0.23	IV
72	Galheta Sul	Brasil	America	0.31	IV

(continued)

No	Coastal site name	Country	Continent	D value	Class
73	Galhetinha	Brasil	America	0.63	III
74	Gamboa Norte	Brasil	America	0.79	II
75	Gamboa Sul	Brasil	America	0.56	III
76	Garopaba	Brasil	America	0.38	IV
77	Gravatá	Brasil	America	0.97	I
78	Guarda do Embaú Norte	Brasil	America	0.61	III
79	Guarda do Embaú Sul	Brasil	America	0.81	II
80	Ibiraquera Centro	Brasil	America	0.51	III
81	Ibiraquera Norte	Brasil	America	0.48	III
82	Ibiraquera Sul	Brasil	America	0.57	III
83	Imbituba	Brasil	America	0.73	II
84	Ipoã Norte	Brasil	America	0.38	IV
85	Ipoã Sul	Brasil	America	0.44	III
86	Itapirubá	Brasil	America	0.15	IV
87	Itapirubá Norte	Brasil	America	0.56	III
88	Itapirubá Sul	Brasil	America	0.22	IV
89	Manelone	Brasil	America	0.95	I
90	Morros dos Conventos Conservada	Brasil	America	0.85	II
91	Nova Camboriú	Brasil	America	0.22	IV
92	Ouvidor	Brasil	America	0.46	III
93	Ouvidor Sul	Brasil	America	0.90	I
94	Paiquerê	Brasil	America	0.29	IV
95	Paraíso	Brasil	America	0.32	IV
96	Passo de Torres	Brasil	America	0.11	IV
97	Passo de Torres Sul	Brasil	America	−0.24	V
98	Portinho do Rosa	Brasil	America	0.64	III
99	Porto de Imbituba	Brasil	America	0.05	IV
100	Praia da Caçamba	Brasil	America	0.22	IV
101	Praia da Ferrugem	Brasil	America	0.68	II
102	Praia da Ponta	Brasil	America	0.58	III
103	Praia da Vila	Brasil	America	0.46	III
104	Praia D'Água	Brasil	America	0.95	I
105	Praia do Gi	Brasil	America	0.43	III
106	Praia do Gi Norte	Brasil	America	0.66	II
107	Praia do Luz	Brasil	America	0.50	III
108	Praia do Mar Grosso	Brasil	America	0.05	IV
109	Praia do Rosa	Brasil	America	0.85	II
110	Praia do Sol Norte	Brasil	America	0.51	III
111	Praia do Sol Sul	Brasil	America	0.14	IV
112	Praia dos Amores	Brasil	America	0.62	III
113	Praia Grande	Brasil	America	0.45	III
114	Praia Vermelha	Brasil	America	0.93	I
115	Prainha	Brasil	America	−0.31	V
116	Ribanceira	Brasil	America	0.36	IV

(continued)

No	Coastal site name	Country	Continent	D value	Class
117	Rincão Barra	Brasil	America	0.04	IV
118	Rosa Sul	Brasil	America	0.56	III
119	Silveira	Brasil	America	0.75	II
120	Siriú Norte	Brasil	America	0.81	II
121	Siriú Sul	Brasil	America	0.91	I
122	Tereza	Brasil	America	0.16	IV
123	Torneiro	Brasil	America	−0.36	V
124	Torneiro Norte	Brasil	America	0.30	IV
125	Vigia	Brasil	America	0.51	III
126	Baía do Sancho	Brasil	America	1.20	I
127	Praia do Leão	Brasil	America	1.15	I
128	Americano	Brasil	America	1.13	I
129	Atalaia/Arenosa	Brasil	America	1.10	I
130	Biboca	Brasil	America	1.09	I
131	Baía dos Porcos	Brasil	America	1.04	I
132	Praia do Bode	Brasil	America	1.01	I
133	Atalaia/Rochosa	Brasil	America	0.99	I
134	Cacimba do Padre	Brasil	America	0.84	II
135	Abreus	Brasil	America	0.79	II
136	Boldró	Brasil	America	0.78	II
137	Praia do Porto/Natural	Brasil	America	0.77	II
138	Praia do Meio	Brasil	America	0.77	II
139	Capim-Açú	Brasil	America	0.74	II
140	Baía do Sueste	Brasil	America	0.71	II
141	Conceição	Brasil	America	0.64	III
142	Caieira	Brasil	America	0.47	III
143	Praia do Cachorro	Brasil	America	0.44	III
144	Praia do Porto/Molhes	Brasil	America	0.36	IV
145	Arrabida	Brasil	America	0.25	IV
146	Armação	Brasil	America	0.48	III
147	Barra da lagoa	Brasil	America	−0.14	V
148	Barra de Ibiraquera	Brasil	America	0.77	II
149	Campeche	Brasil	America	−0.04	V
150	Castelhanos, Brasil	Brasil	America	0.96	I
151	Joaquina	Brasil	America	0.38	IV
152	Matadeiro	Brasil	America	1.01	I
153	Moçambique	Brasil	America	0.78	II
154	Praia Mole	Brasil	America	0.70	II
155	Silveira Sul	Brasil	America	0.91	I
156	No 1 beach Qingdao	China	Asia	0.36	IV
157	Old Stone Man	China	Asia	0.52	III
158	Big Wave bay	China – Hong Kong	Asia	0.96	I

(continued)

No	Coastal site name	Country	Continent	D value	Class
159	Deepwater bay	China – Hong Kong	Asia	0.54	III
160	Repulse bay	China – Hong Kong	Asia	0.04	IV
161	Sheko bay	China – Hong Kong	Asia	0.86	I
162	Stanley	China – Hong Kong	Asia	0.44	III
163	Playa El Convento	Chile	America	0.38	IV
164	Playa Sur Santo Domingo	Chile	America	−0.13	V
165	Las Rocas de Santo Domingo	Chile	America	0.46	III
166	Desembocadura Río Maipo	Chile	America	0.10	IV
167	Caleta San Antonio	Chile	America	−0.94	V
168	Playa Grande Cartagena	Chile	America	−0.48	V
169	Humedal Cartagena	Chile	America	−0.61	V
170	San Sebastian	Chile	America	−0.66	V
171	Playa Chica Las Cruces	Chile	America	−0.15	V
172	El Tabo	Chile	America	−0.37	V
173	Isla Negra	Chile	America	0.89	I
174	Algarrobo	Chile	America	−0.29	V
175	San Alfonso	Chile	America	−0.21	V
176	Humedal San Alfonso	Chile	America	0.75	II
177	Algarrobo	Chile	America	−0.26	V
178	Las Torpederas	Chile	America	−0.55	V
179	Muelle Vergara	Chile	America	−0.50	V
180	Los Marineros	Chile	America	−0.22	V
181	Las Salinas	Chile	America	−0.65	V
182	Reñaca	Chile	America	−0.57	V
183	Montemar	Chile	America	−0.54	V
184	Roca Oceánica	Chile	America	0.95	I
185	La Boca, Desembocadura Río Aconcagua	Chile	America	−0.75	V
186	Mantagua	Chile	America	0.19	IV
187	Ritoque	Chile	America	0.31	IV
188	Ventana	Chile	America	−0.85	V
189	Maitencillo Sur	Chile	America	−0.05	V
190	Maitencillo Centro	Chile	America	−0.48	V
191	Papudo Norte	Chile	America	−0.31	V
192	Papudo Sur	Chile	America	−0.23	V
193	Orongo	Chile	America	1.23	I
194	Rano Kau	Chile	America	1.51	I
195	Ana Kai	Chile	America	1.18	I
196	Hanga Piko	Chile	America	−0.05	V
197	Vate Ote	Chile	America	0.09	IV

(continued)

No	Coastal site name	Country	Continent	D value	Class
198	Playa Pea	Chile	America	−0.03	V
199	Tahai	Chile	America	1.07	I
200	Hanga Kioe	Chile	America	0.71	II
201	Ana Kakenga	Chile	America	1.51	I
202	Ahu Te Peu	Chile	America	1.23	I
203	Anakena	Chile	America	1.24	I
204	Ahu Tongariki	Chile	America	1.56	I
205	Ura Uranga	Chile	America	1.44	I
206	Oroi	Chile	America	1.44	I
207	Muelle Baron	Chile	America	−0.69	V
208	Caleta Portales	Chile	America	−0.85	V
209	Miramar	Chile	America	0.01	IV
210	Caleta Abarca	Chile	America	0.03	IV
211	Vergara	Chile	America	0.20	IV
212	Las Salinas de Pullally	Chile	America	0.39	IV
213	Humedal de Pichicuy	Chile	America	0.48	III
214	Pichicuy	Chile	America	−0.11	V
215	Caleta Pichicuy	Chile	America	−0.26	V
216	La Ballena Sur	Chile	America	0.17	IV
217	La Ballena Norte	Chile	America	−0.04	V
218	Los Molles Sur	Chile	America	−0.09	V
219	Los Molles Centro	Chile	America	−0.42	V
220	Los Molles Norte	Chile	America	−0.76	V
221	Puquen Sur	Chile	America	0.98	I
222	Puquen Norte (Mala Bajada)	Chile	America	0.60	III
223	Zapallar (arriba)	Chile	America	0.48	III
224	Playa Zapallar	Chile	America	0.32	IV
225	Cachagua	Chile	America	0.65	III
226	Laguna de Zapallar	Chile	America	−0.50	V
227	Humedal Laguna de Zapallar	Chile	America	−0.03	V
228	Caleta Horcón	Chile	America	−0.85	V
229	Playa San Mateo	Chile	America	−0.76	V
230	Punta Angeles	Chile	America	−0.04	V
231	Cementerio	Chile	America	−0.27	V
232	Acantilados Santa María	Chile	America	0.10	IV
233	Laguna Verde	Chile	America	0.05	IV
234	Playa Grande Laguna Verde	Chile	America	−0.64	V
235	Quintay Sur	Chile	America	0.40	IV
236	Playa Grande Quintay	Chile	America	0.22	IV
237	Caleta Quintay	Chile	America	−0.50	V
238	Playa Chica Quintay	Chile	America	−0.24	V
239	Duna Tunquen	Chile	America	0.54	III
240	Humedal de Tunquen	Chile	America	0.74	II
241	Balneario Algarrobo Norte	Chile	America	−0.35	V

(continued)

No	Coastal site name	Country	Continent	D value	Class
242	Playa Grande Mirasol	Chile	America	−0.63	V
243	Pescaderia Algarrobo	Chile	America	−0.76	V
244	Playa El Canelo	Chile	America	−0.25	V
245	El Quisco Sur	Chile	America	−0.49	V
246	Punta de Tralca	Chile	America	0.25	IV
247	Playa Punta de Tralca	Chile	America	−0.44	V
248	Estero Cordova	Chile	America	−0.12	V
249	Playa Chepica	Chile	America	−0.67	V
250	Las Cruces Norte	Chile	America	0.44	III
251	Puquen Norte (Límite Regional)	Chile	America	0.82	II
252	Puquen Norte (Arqueología)	Chile	America	1.05	I
253	Las Gaviotas – Punta Curaumilla	Chile	America	0.73	II
254	Playa Las Docas	Chile	America	0.50	III
255	Sapzurro	Colombia	America	0.72	II
256	Capurgana	Colombia	America	0.57	III
257	Aguacate	Colombia	America	0.82	II
258	Acandi	Colombia	America	0.02	IV
259	Acandi	Colombia	America	0.85	II
260	Playona	Colombia	America	1.06	I
261	Trigana	Colombia	America	0.64	III
262	Titumate	Colombia	America	0.53	III
263	Punta yerbasal	Colombia	America	0.46	III
264	Turbo	Colombia	America	−1.11	V
265	Rio turbo	Colombia	America	0.28	IV
266	El totumo	Colombia	America	0.18	IV
267	Necocli	Colombia	America	−0.46	V
268	Rio necocli	Colombia	America	0.11	IV
269	El venado	Colombia	America	0.69	II
270	Zapata	Colombia	America	0.09	IV
271	Damaquiel	Colombia	America	0.53	III
272	Uveros	Colombia	America	−0.27	V
273	San juan de uraba	Colombia	America	0.80	II
274	Arboletes	Colombia	America	−0.28	V
275	Arboletes	Colombia	America	−0.95	V
276	Volcan de lodo	Colombia	America	−0.07	V
277	Los cordobas	Colombia	America	0.63	III
278	Rio cordoba	Colombia	America	0.05	IV
279	San rafael	Colombia	America	0.68	II
280	Rio canalete	Colombia	America	0.51	III
281	Puerto escondido	Colombia	America	0.04	IV
282	Rio cedro	Colombia	America	0.30	IV
283	Broqueles	Colombia	America	0.11	IV
284	Moñitos	Colombia	America	−0.38	V
285	Moñitos	Colombia	America	0.49	III

(continued)

No	Coastal site name	Country	Continent	D value	Class
286	El salvador	Colombia	America	−0.10	V
287	San Bernardo Del Viento	Colombia	America	0.46	III
288	Playa blanca	Colombia	America	−0.35	V
289	El calao	Colombia	America	0.78	II
290	Punta bolivar	Colombia	America	0.09	IV
291	Piedra	Colombia	America	−0.14	V
292	Victoria real	Colombia	America	0.07	IV
293	Tolu	Colombia	America	−0.79	V
294	El frances	Colombia	America	0.19	IV
295	Berrugas	Colombia	America	−0.35	V
296	La cangrejera	Colombia	America	0.79	II
297	Punta seca	Colombia	America	0.84	II
298	Balsillas	Colombia	America	0.59	III
299	El rincon	Colombia	America	−0.66	V
300	Isla palma	Colombia	America	0.57	III
301	Playa blanca	Colombia	America	0.63	III
302	Playa blanca decameron	Colombia	America	−0.12	V
303	Isla bonita	Colombia	America	0.79	II
304	Tierra bomba	Colombia	America	−0.58	V
305	Castillogrande	Colombia	America	−0.33	V
306	Bocagrande	Colombia	America	−0.42	V
307	Marbella	Colombia	America	−0.60	V
308	Las americas	Colombia	America	0.14	IV
309	La boquilla	Colombia	America	−0.62	V
310	Manzanillo	Colombia	America	−0.07	V
311	Manzanillo del mar	Colombia	America	0.45	III
312	Bocacanoas	Colombia	America	0.29	IV
313	Arroyo de piedras	Colombia	America	0.49	III
314	Galerazamba	Colombia	America	0.23	IV
315	Lomita de arena	Colombia	America	0.36	IV
316	Punta canoas	Colombia	America	0.65	II
317	Punta roca	Colombia	America	−0.61	V
318	Salgar	Colombia	America	−0.60	V
319	Pradomar – resort	Colombia	America	−1.01	V
320	Pradomar – urbana	Colombia	America	−1.23	V
321	Puerto colombia – norte	Colombia	America	−1.25	V
322	Puerto colombia – malecon	Colombia	America	−1.22	V
323	Puerto velero – expuesto	Colombia	America	−0.23	V
324	Puerto velero – resort	Colombia	America	0.06	IV
325	Puerto velero – punta velero	Colombia	America	−0.22	V
326	Puerto velero – Mirador de velero	Colombia	America	0.20	IV
327	Caño dulce	Colombia	America	−0.32	V
328	Puerto caiman	Colombia	America	−0.10	V
329	Playa mendoza	Colombia	America	0.19	IV

(continued)

No	Coastal site name	Country	Continent	D value	Class
330	Turipana	Colombia	America	0.02	IV
331	Tubara	Colombia	America	−0.19	V
332	Palmarito	Colombia	America	0.29	IV
333	Playa linda	Colombia	America	0.05	IV
334	Santa Verónica – Cajacopi	Colombia	America	−0.51	V
335	Santa Verónica	Colombia	America	−0.93	V
336	Salinas del rey	Colombia	America	0.05	IV
337	Loma de piedra	Colombia	America	−0.08	V
338	Aguamarina	Colombia	America	0.10	IV
339	Bocatocinos	Colombia	America	0.05	IV
340	Punta astilleros	Colombia	America	−0.08	V
341	Salinas de galerazamba	Colombia	America	−0.53	V
342	Galerazamba	Colombia	America	−0.65	V
343	Salamanca	Colombia	America	0.77	II
344	Tasajera	Colombia	America	0.73	II
345	Pueblo viejo	Colombia	America	0.71	II
346	Cienaga	Colombia	America	0.72	II
347	Villa tanga	Colombia	America	0.31	IV
348	Aeropuerto	Colombia	America	−0.33	V
349	Irotama	Colombia	America	−0.02	V
350	Pozos colorados	Colombia	America	0.13	IV
351	Rodadero	Colombia	America	0.04	IV
352	Gaira	Colombia	America	0.25	IV
353	Playa lipe	Colombia	America	0.88	I
354	Los cocos	Colombia	America	−0.46	V
355	Santa Marta	Colombia	America	−0.48	V
356	Taganga	Colombia	America	−0.30	V
357	Playa brava	Colombia	America	1.12	I
358	Bahia concha	Colombia	America	1.08	I
359	Macuaca	Colombia	America	1.36	I
360	Chengue	Colombia	America	1.29	I
361	Gayraca	Colombia	America	0.96	I
362	7 olas	Colombia	America	1.18	I
363	Cinto	Colombia	America	1.14	I
364	Neguanje	Colombia	America	1.14	I
365	Playa del cabo	Colombia	America	1.17	I
366	San juan de guia	Colombia	America	0.98	I
367	La piscina	Colombia	America	0.97	I
368	Playa arenita	Colombia	America	0.86	I
369	Paraiso	Colombia	America	0.93	I
370	Arrecifes	Colombia	America	1.16	I
371	Cañaveral	Colombia	America	1.03	I
372	Castilletes	Colombia	America	1.07	I
373	Rio piedras	Colombia	America	0.85	II

(continued)

No	Coastal site name	Country	Continent	D value	Class
374	Los naranjos	Colombia	America	0.81	II
375	Mendihuaca	Colombia	America	0.41	III
376	Guachaca	Colombia	America	0.41	III
377	Quebrada valencia	Colombia	America	0.58	III
378	Buritaca	Colombia	America	0.11	IV
379	Don diego	Colombia	America	0.59	III
380	Perico	Colombia	America	0.59	III
381	Los muchachitos	Colombia	America	0.40	III
382	Marquetalia	Colombia	America	0.56	III
383	Palomino	Colombia	America	0.39	IV
384	Repunton grande	Colombia	America	0.56	III
385	Playa de los holandeses	Colombia	America	0.68	II
386	Termoelectrica	Colombia	America	0.07	IV
387	Dibulla	Colombia	America	0.65	II
388	Camarones	Colombia	America	0.69	II
389	Riohacha sur	Colombia	America	−0.85	V
390	Riohacha norte	Colombia	America	−0.34	V
391	Valle de los cangrejos	Colombia	America	0.52	III
392	Mayapo	Colombia	America	0.67	II
393	Manaure viejo	Colombia	America	0.19	IV
394	Manaure nuevo	Colombia	America	−0.20	V
395	Carrizal	Colombia	America	0.62	III
396	Cabo de la vela sur	Colombia	America	1.06	I
397	Cabo de la vela norte	Colombia	America	1.15	I
398	Bahia honda	Colombia	America	1.09	I
399	Punta gallinas	Colombia	America	1.20	I
400	Puerto lopez	Colombia	America	1.10	I
401	Castilletes	Colombia	America	1.13	I
402	Cocoplum	Colombia	America	0.77	II
403	Manzanillo	Colombia	America	1.02	I
404	San andres	Colombia	America	0.07	IV
405	Muri beach	Cook islands	Oceania	1.16	I
406	Zlatni Rat	Croatia	Europe	1.21	I
407	Zaglav	Croatia	Europe	1.18	I
408	Biglake	Croatia	Europe	1.18	I
409	Struga	Croatia	Europe	1.08	I
410	Zabarje	Croatia	Europe	1.07	I
411	Sv Mihajho	Croatia	Europe	1.06	I
412	Mesuporat	Croatia	Europe	0.83	II
413	Zaklopatica	Croatia	Europe	0.82	II
414	Korchula	Croatia	Europe	0.80	II
415	Polace	Croatia	Europe	0.78	II
416	Srebrna	Croatia	Europe	0.69	II
417	Stiniva	Croatia	Europe	0.68	II

(continued)

No	Coastal site name	Country	Continent	D value	Class
418	Salbunara	Croatia	Europe	0.67	II
419	Skrivina Lika	Croatia	Europe	0.67	II
420	Sobre	Croatia	Europe	0.66	II
421	SlariGrad	Croatia	Europe	0.66	II
422	Porat	Croatia	Europe	0.65	II
423	Komiza 2	Croatia	Europe	0.54	III
424	Komiza 1	Croatia	Europe	0.51	III
425	Milna	Croatia	Europe	0.49	III
426	Croatia Beach, Cavtat	Croatia	Europe	0.47	III
427	Jadran	Croatia	Europe	0.46	III
428	Sustipan	Croatia	Europe	0.44	III
429	Ploce	Croatia	Europe	0.40	III
430	Bonje	Croatia	Europe	0.30	IV
431	Bobovisca Luka	Croatia	Europe	0.26	IV
432	Albalros sbeach, Cavlal	Croatia	Europe	0.25	IV
433	Dance	Croatia	Europe	0.25	IV
434	Kastelet	Croatia	Europe	0.22	IV
435	Milna	Croatia	Europe	0.21	IV
436	Supetar	Croatia	Europe	0.13	IV
437	Uvala Lapad	Croatia	Europe	−0.26	V
438	Bacvice	Croatia	Europe	−0.26	V
439	Puerto Esperanza	Cuba	America	0.06	IV
440	Cayo Levisa	Cuba	America	0.93	I
441	Playa El Morrillo	Cuba	America	0.10	IV
442	La Altura	Cuba	America	0.56	III
443	Playa Carenero	Cuba	America	0.35	IV
444	Playa Banes	Cuba	America	0.32	IV
445	Bocaciega	Cuba	America	0.70	II
446	Guanabo	Cuba	America	0.60	III
447	La Puntilla	Cuba	America	1.14	I
448	Jibacoa	Cuba	America	0.58	III
449	Los Cocos	Cuba	America	0.93	I
450	Arroyo Berbejo	Cuba	America	0.87	I
451	El Judío	Cuba	America	−0.90	V
452	Tenis	Cuba	America	−0.87	V
453	Allende	Cuba	America	−0.22	V
454	El Mamey	Cuba	America	0.64	III
455	Bueyvaca	Cuba	America	0.57	III
456	Faro Maya	Cuba	America	−0.42	V
457	Canal Oeste	Cuba	America	0.23	IV
458	Canal Este	Cuba	America	0.12	IV
459	Calle 46	Cuba	America	0.80	II
460	Calle 48–50	Cuba	America	0.54	III
461	Calle 55	Cuba	America	0.80	II

(continued)

No	Coastal site name	Country	Continent	D value	Class
462	Calle 57 Museo	Cuba	America	0.79	II
463	Plaza America	Cuba	America	0.38	IV
464	Arenas Blancas	Cuba	America	0.41	III
465	Las Américas	Cuba	America	0.47	III
466	Sandals	Cuba	America	0.80	II
467	Solymar	Cuba	America	0.06	IV
468	Barlovento	Cuba	America	0.38	IV
469	Brisas	Cuba	America	0.69	II
470	Playa El Salto	Cuba	America	0.58	III
471	La Panchita	Cuba	America	0.05	IV
472	Isabela	Cuba	America	−1.08	V
473	La Punta	Cuba	America	0.18	IV
474	Punta Periquillo	Cuba	America	0.75	II
475	Playa Cayo Ensenacho	Cuba	America	0.43	III
476	Hotel Meliá (CSM)	Cuba	America	0.41	III
477	Las Gaviotas	Cuba	America	0.81	II
478	Playa Pilar	Cuba	America	0.76	II
479	Hotel Meliá (CG)	Cuba	America	0.76	II
480	Campismo Uva Calita	Cuba	America	0.62	III
481	Playa Larga (CA)	Cuba	America	0.76	II
482	Hotel Sol Cayo Coco	Cuba	America	0.72	II
483	Playa Piloto	Cuba	America	−0.30	V
484	S. Lucia La Boca	Cuba	America	0.40	III
485	S. Lucia Playa Coco	Cuba	America	0.03	IV
486	S. Lucia Natural	Cuba	America	0.58	III
487	S. Lucia Tararaco	Cuba	America	0.74	II
488	S. Lucia La Concha	Cuba	America	0.29	IV
489	Playa del Puerco	Cuba	America	0.48	III
490	Los Bajos	Cuba	America	0.31	IV
491	Playa Blanca (H)	Cuba	America	0.66	II
492	Don Lino	Cuba	America	0.79	II
493	Playa Pesquero	Cuba	America	0.81	II
494	Guardalavaca	Cuba	America	0.26	IV
495	Puerto Rico	Cuba	America	0.43	III
496	Morales Laguna	Cuba	America	0.15	IV
497	Morales Village	Cuba	America	−0.09	V
498	Juan Vicente 1	Cuba	America	−0.53	V
499	Juan Vicente 2	Cuba	America	−0.29	V
500	Corinthia	Cuba	America	0.54	III
501	Mejia	Cuba	America	0.02	IV
502	Maguana	Cuba	America	0.47	III
503	Toa	Cuba	America	0.21	IV
504	Duaba	Cuba	America	0.51	III
505	Cajuajo	Cuba	America	0.42	III

(continued)

No	Coastal site name	Country	Continent	D value	Class
506	Playa Bariguita	Cuba	America	0.92	I
507	Cajobabo	Cuba	America	0.68	II
508	Imias	Cuba	America	0.51	III
509	Baconao	Cuba	America	0.80	II
510	Cazonal	Cuba	America	0.47	III
511	Sigua	Cuba	America	0.29	IV
512	Verraco	Cuba	America	0.40	III
513	Juragua	Cuba	America	0.31	IV
514	Siboney	Cuba	America	0.29	IV
515	La Estrella	Cuba	America	0.39	IV
516	Mar Verde	Cuba	America	−0.19	V
517	Buey Cabon	Cuba	America	−0.07	V
518	Caletón Blanco	Cuba	America	−0.02	V
519	Hicacal	Cuba	America	0.73	II
520	El Frances	Cuba	America	−0.10	V
521	Playa Colorada	Cuba	America	0.92	I
522	Sierra Mar	Cuba	America	0.38	IV
523	Playa Blanca (S)	Cuba	America	1.03	I
524	Chivirico	Cuba	America	0.04	IV
525	Las Coloradas	Cuba	America	0.37	IV
526	Playa Carenero	Cuba	America	0.51	III
527	Playa Levisa	Cuba	America	0.62	III
528	Ancón	Cuba	America	0.48	III
529	La Boca (SS)	Cuba	America	−0.24	V
530	Yaguanabo	Cuba	America	0.22	IV
531	Playa Ingles	Cuba	America	0.04	IV
532	El Río	Cuba	America	0.34	IV
533	Campamento	Cuba	America	0.20	IV
534	Playa Larga (M)	Cuba	America	0.08	IV
535	Buenaventura	Cuba	America	0.02	IV
536	Playa Sirena	Cuba	America	0.86	I
537	Playa Rosario	Cuba	America	0.26	IV
538	Playa Mayabeque	Cuba	America	−0.48	V
539	Playa Majana	Cuba	America	−0.09	V
540	Aligade, N Cyprus,	Cyprus	Europe	0.57	III
541	Playa Tarqui	Ecuador	America	−0.41	V
542	Playa Piedra Larga	Ecuador	America	−0.99	V
543	Playa Jamabeli	Ecuador	America	−0.68	V
544	Playa Manta	Ecuador	America	0.57	III
545	Makardi Bay	Egypt	Africa	−0.02	V
546	Modriki Island	Fiji	Oceania	1.21	I
547	Paradise Island	Fiji	Oceania	1.17	I
548	Dyrhataeg	Iceland	Europe	0.59	III
549	Retynisfjara	Iceland	Europe	1.11	I

(continued)

No	Coastal site name	Country	Continent	D value	Class
550	Glacial Lagoon	Iceland	Europe	0.63	III
551	Hestgeral	Iceland	Europe	0.44	III
552	Hefn	Iceland	Europe	−0.06	V
553	Bvottarskriodr	Iceland	Europe	0.56	III
554	Bulandstindur	Iceland	Europe	0.87	I
555	Reykir	Iceland	Europe	0.53	III
556	Land Ranger	Iceland	Europe	0.71	II
557	Djupalonssandur	Iceland	Europe	0.92	I
558	Hellner	Iceland	Europe	0.71	II
559	Oseyares	Iceland	Europe	0.79	II
560	Taj hotel Goa India	India	Asia	0.55	III
561	Fort, Goa, India	India	Asia	0.59	III
562	Candolim Beach	India	Asia	0.47	III
563	River Princess Site	India	Asia	0.48	III
564	Giants Causeway	Ireland	Europe	0.87	I
565	Doonbeg	Ireland	Europe	0.90	I
566	Ryans Daughter Area	Ireland	Europe	0.86	I
567	Mather Cliffs	Ireland	Europe	0.76	II
568	Burren Area	Ireland	Europe	0.26	IV
569	Inch	Ireland	Europe	−0.26	V
570	Strangford Loch	Ireland	Europe	−0.56	V
571	Bombarse, Sardinia	Italy	Europe	0.16	IV
572	Pelosa, Sardinia	Italy	Europe	0.84	II
573	Acqua Calde, Lipari	Italy – Aeolian Isles	Europe	0.26	IV
574	Havana beach Salina	Italy – Aeolian Isles	Europe	0.62	III
575	Il Positano beach, Salina	Italy – Aeolian Isles	Europe	0.88	I
576	Vulcano, Vuklcano	Italy – Aeolian Isles	Europe	0.55	III
577	Spiaggia delle Graticciate	Italy	Europe	0.14	IV
578	Spiaggia delle Murelle	Italy	Europe	−0.39	V
579	Spiaggia Spinicci	Italy	Europe	−0.22	V
580	Spiaggia Pian di Spille	Italy	Europe	0.08	IV
581	Lido di Tarquinia	Italy	Europe	−0.95	V
582	Spiaggia dei Bagni di Sant'Agostino	Italy	Europe	0.12	IV
583	Spiaggia di Santa Marinella	Italy	Europe	−0.33	V
584	Spiaggia di Santa Severa	Italy	Europe	0.34	IV
585	Spiaggia Torre Flavia	Italy	Europe	0.59	III
586	Lido di Maccaresse	Italy	Europe	0.22	IV
587	Spiaggia del Controvento	Italy	Europe	0.20	IV
588	Spiaggia Settimo Cielo	Italy	Europe	0.26	IV
589	Spiaggia Oasi Naturista Capoccotta	Italy	Europe	0.26	IV
590	Lido dei Gigli	Italy	Europe	0.19	IV
591	Spiaggia Tor Caldara	Italy	Europe	−0.08	V
592	Spiaggia Grotte di Nerone	Italy	Europe	−0.13	V
593	Lido di Foce Verde	Italy	Europe	−0.39	V

(continued)

No	Coastal site name	Country	Continent	D value	Class
594	Lido di Capo Portiere	Italy	Europe	0.24	IV
595	Spiaggia La Bufalara	Italy	Europe	0.65	II
596	Spiaggia Dune di Sabaudia	Italy	Europe	0.21	IV
597	Spiaggia di San Felice Circeo	Italy	Europe	0.01	IV
598	Spiaggia dell'Angolo	Italy	Europe	1.03	I
599	Lido Costadoro	Italy	Europe	0.50	III
600	Spiaggia dei 300 Gradini	Italy	Europe	0.69	II
601	Spiaggia dell'Ariana	Italy	Europe	−0.09	V
602	Spiaggia Fontania	Italy	Europe	−0.07	V
603	Spiaggia Serapo	Italy	Europe	−0.16	V
604	Spiaggia di Vincidio	Italy	Europe	−0.16	V
605	Lido Lo Scoglio	Italy	Europe	−0.37	V
606	Caucana	Italy	Europe	0.58	III
607	Spiaggia dei Sassolini	Italy	Europe	−0.11	V
608	Irohzaki Lighthouse	Japan	Asia	1.31	I
609	Ebisu Beach	Japan	Asia	1.13	I
610	Tojo Beach	Japan	Asia	0.73	II
611	Irita	Japan	Asia	0.41	III
612	South Beach	Jordan	Asia	0.38	IV
613	Central Castle Beach	Jordan	Asia	0.32	IV
614	South Beach Public	Jordan	Asia	0.22	IV
615	Barracuda Beach	Jordan	Asia	−0.46	V
616	Japanese Garden Beach	Jordan	Asia	−0.40	V
617	Movenpick Resort Beach	Jordan	Asia	0.60	III
618	Paradise Island	Maldives	Oceania	1.10	I
619	Artificial Beach, Male	Maldives	Oceania	0.14	IV
620	San Blas beach	Malta	Europe	1.10	I
621	Dingli Cliffs	Malta	Europe	0.97	I
622	Crystal Bay	Malta	Europe	0.94	I
623	Fungus Rock	Malta	Europe	0.77	II
624	Blue Lagoon	Malta	Europe	0.72	II
625	Santa Maria Bay	Malta	Europe	0.72	II
626	Azure Window	Malta	Europe	0.67	II
627	Tat – Tocc	Malta	Europe	0.65	II
628	Ta' Cenc	Malta	Europe	0.64	III
629	Hekka point	Malta	Europe	0.60	III
630	Ghajn Tuffieha	Malta	Europe	0.56	III
631	Manikata	Malta	Europe	0.56	III
632	Mgarr ix-Xini	Malta	Europe	0.54	III
633	Ramla	Malta	Europe	0.52	III
634	Calypso's Cave	Malta	Europe	0.48	III
635	Xatt l-Ahmar	Malta	Europe	0.42	III
636	Wied il-Ghasri	Malta	Europe	0.41	III
637	Ta' Xilep	Malta	Europe	0.40	III

(continued)

No	Coastal site name	Country	Continent	D value	Class
638	Sea Cave	Malta	Europe	0.38	IV
639	Mellieha	Malta	Europe	0.37	IV
640	Hondoq ir-Rummien	Malta	Europe	0.25	IV
641	Wied il-Mielah	Malta	Europe	0.17	IV
642	Kercem Cliffs	Malta	Europe	0.16	IV
643	Xlendi tower	Malta	Europe	0.13	IV
644	Iz-Zewwieqa	Malta	Europe	0.12	IV
645	White Towers	Malta	Europe	0.10	IV
646	Xwieni Point	Malta	Europe	0.08	IV
647	Inland Sea	Malta	Europe	0.07	IV
648	Quala	Malta	Europe	−0.07	V
649	Ghallis Rocks coastline	Malta	Europe	−0.12	V
650	Xlendi bay	Malta	Europe	−0.13	V
651	Marsalforn	Malta	Europe	−0.37	V
652	Bahar Ic-caghaq	Malta	Europe	−0.41	V
653	St. George's Bay	Malta	Europe	−0.64	V
654	Kamkoum El Baz	Morocco	Africa	0.92	I
655	Plage Rouge	Morocco	Africa	0.66	II
656	Sidi Driss	Morocco	Africa	0.61	III
657	Taourirt	Morocco	Africa	0.77	II
658	Boukana East	Morocco	Africa	0.44	III
659	Chemaala	Morocco	Africa	0.54	III
660	Firma	Morocco	Africa	0.54	III
661	Sidi Hsaine	Morocco	Africa	0.55	III
662	Sidi Lahcen	Morocco	Africa	0.47	III
663	Tazaghine	Morocco	Africa	0.43	III
664	Tibouda	Morocco	Africa	0.48	III
665	Kalat	Morocco	Africa	0.40	III
666	Arekmane	Morocco	Africa	0.14	IV
667	Amejaou	Morocco	Africa	0.36	IV
668	BoukanaWest	Morocco	Africa	0.27	IV
669	Charranna	Morocco	Africa	0.36	IV
670	Miami West	Morocco	Africa	0.09	IV
671	Miami East	Morocco	Africa	0.09	IV
672	Ras El Ma	Morocco	Africa	0.08	IV
673	Sidi Abderrazak	Morocco	Africa	0.33	IV
674	Mdiq	Morocco	Africa	−0.23	V
675	Souani	Morocco	Africa	0.09	IV
676	Kaikura Seal Colony	New Zealand	Oceania	1.13	I
677	Punakaiki	New Zealand	Oceania	1.06	I
678	Karekare	New Zealand	Oceania	1.29	I
679	Sumner	New Zealand	Oceania	1.17	I
680	Piha	New Zealand	Oceania	0.97	I
681	Taylors Mistake	New Zealand	Oceania	0.91	I

(continued)

No	Coastal site name	Country	Continent	D value	Class
682	Abel Taman,	New Zealand	Oceania	0.88	I
683	Kaik Peninsula	New Zealand	Oceania	1.37	I
684	Nape Nape	New Zealand	Oceania	1.17	I
685	Hurunui	New Zealand	Oceania	1.04	I
686	Seal Colony	New Zealand	Oceania	1.04	I
687	Motunau	New Zealand	Oceania	1.00	I
688	Waikku Beach North	New Zealand	Oceania	0.90	I
689	Okainis Bay	New Zealand	Oceania	0.86	I
690	WadiNimrin	New Zealand	Oceania	0.84	II
691	Waikuku Beach	New Zealand	Oceania	0.84	II
692	Woodend Beach	New Zealand	Oceania	0.81	II
693	Leihfield Beach	New Zealand	Oceania	0.81	II
694	Corsay Bay	New Zealand	Oceania	0.81	II
695	Purau	New Zealand	Oceania	0.78	II
696	Claverly	New Zealand	Oceania	0.77	II
697	S-South Bay	New Zealand	Oceania	0.76	II
698	French Farm	New Zealand	Oceania	0.75	II
699	Pines Beach	New Zealand	Oceania	0.70	II
700	Gore Bay	New Zealand	Oceania	0.68	II
701	Oaro	New Zealand	Oceania	0.65	II
702	Akaroa	New Zealand	Oceania	0.61	III
703	Pingeon Bay	New Zealand	Oceania	0.58	III
704	Field station	New Zealand	Oceania	0.57	III
705	Goose bay	New Zealand	Oceania	0.55	III
706	Amberly Beach	New Zealand	Oceania	0.54	III
707	South Bay	New Zealand	Oceania	0.53	III
708	Diamond Harbour	New Zealand	Oceania	0.50	III
709	N-South Bay	New Zealand	Oceania	0.48	III
710	Port Levy	New Zealand	Oceania	0.45	III
711	South Bay Town	New Zealand	Oceania	0.42	III
712	Birdlings Flat	New Zealand	Oceania	0.39	IV
713	Armers Beach	New Zealand	Oceania	0.30	IV
714	Kaikoura Town	New Zealand	Oceania	0.28	IV
715	Waimari	New Zealand	Oceania	0.24	IV
716	New Brighton	New Zealand	Oceania	0.05	IV
717	Harjani	Pakistan	Asia	0.49	III
718	Jiwani	Pakistan	Asia	1.14	I
719	Kaka Pir	Pakistan	Asia	0.50	III
720	Keti Bandar	Pakistan	Asia	0.41	III
721	Miani Hor	Pakistan	Asia	1.09	I
722	Mubarak	Pakistan	Asia	0.94	I
723	La Playa	Peru	America	0.16	IV
724	Swinoujscie	Poland	Europe	0.57	III
725	Swinoujscie W	Poland	Europe	1.05	I

(continued)

No	Coastal site name	Country	Continent	D value	Class
726	Miedzyzdroje dune	Poland	Europe	0.11	IV
727	Miedzyzdroje cliff	Poland	Europe	0.71	II
728	Miedzyzdroje Amber Balti	Poland	Europe	−0.04	V
729	Gosan Hill	Poland	Europe	1.01	I
730	Miedzywodzie W	Poland	Europe	0.96	I
731	Dziwnow seawall	Poland	Europe	0.25	IV
732	Trzesacz	Poland	Europe	0.31	IV
733	Kolobrzeg Faro	Poland	Europe	0.03	IV
734	Smoldzino fossil forest	Poland	Europe	1.09	I
735	Smoldzino beach	Poland	Europe	1.04	I
736	Karwia	Poland	Europe	0.76	II
737	Jastrzebia Gora seawall	Poland	Europe	0.68	II
738	Jastrzebia Gora Titanic	Poland	Europe	0.51	III
739	Smoldzino fossil forest	Poland	Europe	1.09	I
740	Swinoujscie W	Poland	Europe	1.05	I
741	Smoldzino beach	Poland	Europe	1.04	I
742	Gosan Hill	Poland	Europe	1.01	I
743	Miedzywodzie W	Poland	Europe	0.96	I
744	Karwia	Poland	Europe	0.76	II
745	Miedzyzdroje cliff	Poland	Europe	0.71	II
746	Jastrzebia Gora seawall	Poland	Europe	0.68	II
747	Swinoujscie	Poland	Europe	0.57	III
748	Jastrzebia Gora Titanic	Poland	Europe	0.51	III
749	Trzesacz	Poland	Europe	0.31	IV
750	Dziwnow seawall	Poland	Europe	0.25	IV
751	Miedzyzdroje dune	Poland	Europe	0.11	IV
752	Kolobrzeg Faro	Poland	Europe	0.03	IV
753	Miedzyzdroje Amber Balti	Poland	Europe	−0.04	V
754	Ursa	Portugal	Europe	1.24	I
755	Cabo Roco	Portugal	Europe	1.16	I
756	Praia do Rei	Portugal	Europe	0.92	I
757	Espichel	Portugal	Europe	0.81	II
758	Arrabida	Portugal	Europe	0.78	II
759	Guincho	Portugal	Europe	0.69	II
760	Praia Adraga	Portugal	Europe	0.61	III
761	Lagoa de Albufeira	Portugal	Europe	0.54	III
762	Praia das Macas	Portugal	Europe	0.54	III
763	Lizandro	Portugal	Europe	0.49	III
764	Magoito	Portugal	Europe	0.48	III
765	Praia Grande	Portugal	Europe	0.42	III
766	Fonte de Telha	Portugal	Europe	0.41	III
767	Sesimbra	Portugal	Europe	0.35	IV
768	Setubal	Portugal	Europe	0.30	IV
769	Ribeira d'Ilhas	Portugal	Europe	0.28	IV

(continued)

No	Coastal site name	Country	Continent	D value	Class
770	Cascais	Portugal	Europe	0.04	IV
771	Costa da Caparica	Portugal	Europe	0.00	V
772	Caxias	Portugal	Europe	0.02	IV
773	Carcavelos	Portugal	Europe	0.00	V
774	Belem	Portugal	Europe	−0.29	V
775	Ericeira	Portugal	Europe	−0.55	V
776	Praia dos Mosteiros	Portugal – Azores	Europe	0.30	IV
777	Piscina das Feteiras	Portugal – Azores	Europe	0.61	III
778	Praia do Pópulo	Portugal – Azores	Europe	0.26	IV
779	Praia das Milícias	Portugal – Azores	Europe	0.13	IV
780	Praia dos Areais	Portugal – Azores	Europe	0.47	III
781	Piscinas São Vicente	Portugal – Azores	Europe	0.31	IV
782	Praia dos Moinhos	Portugal – Azores	Europe	0.80	II
783	Porto da Caloura	Portugal – Azores	Europe	0.53	III
784	Praia de Agua d'Alto	Portugal – Azores	Europe	0.26	IV
785	Praia da Ribeira Quente	Portugal – Azores	Europe	0.28	IV
786	Praia da Povoação	Portugal – Azores	Europe	0.37	IV
787	El Rompidillo	Spain	Europe	−0.48	V
788	Santa María del Mar	Spain	Europe	−0.70	V
789	Santa Amalia	Spain	Europe	−0.44	V
790	La Victoria	Spain	Europe	−0.36	V
791	La Puntilla	Spain	Europe	−0.36	V
792	La Chucha 2	Spain	Europe	−0.28	V
793	El Rinconcillo	Spain	Europe	−0.24	V
794	Fuentebravia	Spain	Europe	−0.21	V
795	La Costilla	Spain	Europe	−0.21	V
796	Las Acacias	Spain	Europe	−0.14	V
797	La Caleta	Spain	Europe	−0.05	V
798	Benhajarafe	Spain	Europe	0.03	IV
799	La Barrosa	Spain	Europe	0.03	IV
800	Torreguadiaro	Spain	Europe	0.14	IV
801	Torre del Mar	Spain	Europe	0.21	IV
802	La Chucha 1	Spain	Europe	0.23	IV
803	Los Alamos	Spain	Europe	0.23	IV
804	Fontanilla	Spain	Europe	0.24	IV
805	Palmones	Spain	Europe	0.29	IV
806	Santa Barbara	Spain	Europe	0.30	IV
807	Peñoncillo	Spain	Europe	0.38	IV
808	El Rinconcillo	Spain	Europe	0.41	III
809	Islantilla	Spain	Europe	0.42	III
810	La Rada	Spain	Europe	0.48	III
811	Isdabe (Casasola)	Spain	Europe	0.49	III
812	La Guardia	Spain	Europe	0.51	III
813	La Cortadura	Spain	Europe	0.51	III

(continued)

No	Coastal site name	Country	Continent	D value	Class
814	La Atunara	Spain	Europe	0.54	III
815	Las Negras	Spain	Europe	0.57	III
816	Nagueles	Spain	Europe	0.57	III
817	Conil	Spain	Europe	0.64	III
818	Almayate	Spain	Europe	0.66	II
819	Las Alberquillas	Spain	Europe	0.67	II
820	Rijana	Spain	Europe	0.69	II
821	La Cala del Aceite	Spain	Europe	0.73	II
822	Nueva Andalucia	Spain	Europe	0.75	II
823	Valdevaqueros	Spain	Europe	0.82	II
824	Punta Umbría	Spain	Europe	0.83	II
825	Base Arenosillo	Spain	Europe	0.87	I
826	Cala del Carnaje	Spain	Europe	0.88	I
827	La Barrosa (Hotel)	Spain	Europe	0.89	I
828	Natural Park of Doñana	Spain	Europe	0.91	I
829	Calas de Roche	Spain	Europe	0.99	I
830	Playa de Monsul	Spain	Europe	1.13	I
831	Playa de los Genoveses	Spain	Europe	1.26	I
832	La Arena	Spain	Europe	0.47	III
833	Las Arenas	Spain	Europe	−0.73	V
834	Arrigunaga	Spain	Europe	0.15	IV
835	Gorrondatxe	Spain	Europe	0.76	II
836	Barinatxe	Spain	Europe	0.85	I
837	Atxabiribil	Spain	Europe	0.58	III
838	Plentzia	Spain	Europe	0.20	IV
839	Gorliz	Spain	Europe	0.51	III
840	Bakio	Spain	Europe	0.15	IV
841	Aritzatxu	Spain	Europe	0.13	IV
842	Laidatxu	Spain	Europe	−0.41	V
843	Laida	Spain	Europe	0.54	III
844	Laga	Spain	Europe	0.96	I
845	Ea	Spain	Europe	−0.13	V
846	Ogeia	Spain	Europe	0.54	III
847	Karraspio	Spain	Europe	0.68	II
848	Saturraran	Spain	Europe	0.71	II
849	Deba	Spain	Europe	0.31	IV
850	Itzurun	Spain	Europe	0.52	III
851	Santiago	Spain	Europe	0.54	III
852	Malkorbe	Spain	Europe	0.05	IV
853	Zarautz	Spain	Europe	0.07	IV
854	Antilla	Spain	Europe	0.42	III
855	Ondarreta	Spain	Europe	0.22	IV
856	La Concha	Spain	Europe	0.27	IV
857	Zurriola	Spain	Europe	0.00	IV

(continued)

No	Coastal site name	Country	Continent	D value	Class
858	Hondarribia	Spain	Europe	0.13	IV
859	Platja Canyerets	Spain	Europe	0.56	III
860	Cala Salionç	Spain	Europe	0.10	IV
861	Cala Giverola	Spain	Europe	0.52	III
862	Tossa	Spain	Europe	0.11	IV
863	Cala Roca Grossa	Spain	Europe	0.50	III
864	Pla de l'ós	Spain	Europe	−0.50	V
865	Ocata	Spain	Europe	−0.60	V
866	Montgat	Spain	Europe	−0.37	V
867	Mar Bella	Spain	Europe	−0.52	V
868	Bogatell	Spain	Europe	−0.57	V
869	Nova Icària	Spain	Europe	−0.58	V
870	Somorrostro	Spain	Europe	−0.61	V
871	La Barceloneta	Spain	Europe	−0.62	V
872	Sant Miquel	Spain	Europe	−0.62	V
873	San Sebastià	Spain	Europe	−0.62	V
874	Platges de Gavà	Spain	Europe	0.28	IV
875	Castelldefels	Spain	Europe	−0.24	V
876	Port ginesta	Spain	Europe	−0.38	V
877	Cala Vallcarca	Spain	Europe	−0.34	V
878	Sitges	Spain	Europe	−0.40	V
879	Platja dels Capellans	Spain	Europe	0.50	III
880	Cala Vidre	Spain	Europe	0.40	III
881	L'atmella	Spain	Europe	0.48	III
882	Terme del Perelló	Spain	Europe	0.53	III
883	Cap Roig	Spain	Europe	0.52	III
884	L'ampolla	Spain	Europe	0.10	IV
885	Platja Trabucador	Spain	Europe	0.79	II
886	Lagonaaria	Tahiti	Oceania	0.95	I
887	Trou de Souffleur	Tahiti	Oceania	0.85	II
888	Plage de Surf	Tahiti	Oceania	0.71	II
889	Çıralı Mid-section	Turkey	Asia	1.31	I
890	Çıralı Karaburun	Turkey	Asia	1.26	I
891	Phasalis Small Bay	Turkey	Asia	1.08	I
892	Phaselis Large Bay	Turkey	Asia	0.91	I
893	Tisan Back Bay Mersin	Turkey	Asia	0.83	II
894	Tisan Tample, Mersin	Turkey	Asia	0.68	II
895	Karaburun Akyar Mersin	Turkey	Asia	0.67	II
896	Göksu Hurma, Mersin	Turkey	Asia	0.61	III
897	Alata West, Mersin	Turkey	Asia	0.31	IV
898	Alata Mid, Mersin	Turkey	Asia	0.29	IV
899	Antalya Old Harbour	Turkey	Asia	0.19	IV
900	Tekirova North	Turkey	Asia	0.19	IV
901	Tekirova South	Turkey	Asia	0.18	IV

(continued)

No	Coastal site name	Country	Continent	D value	Class
902	Konyaaltı West	Turkey	Asia	0.10	IV
903	Konyaaltı East	Turkey	Asia	0.09	IV
904	Alata East, Mersin	Turkey	Asia	0.07	IV
905	Konyaaltı Middle	Turkey	Asia	0.04	IV
906	Antalya Waterfalls	Turkey	Asia	−0.01	V
907	Antalya Lara Barınak	Turkey	Asia	−0.16	V
908	Antalya Dedeman Hotel	Turkey	Asia	−0.21	V
909	Lara Beach	Turkey	Asia	−0.28	V
910	Kız Kalesi Mersin	Turkey	Asia	−0.58	V
911	Little Haven, Wales	United Kingdom	Europe	1.00	I
912	Poppit, Wales	United Kingdom	Europe	0.89	I
913	Nash Point, Wales	United Kingdom	Europe	0.74	II
914	Mwnt, Wales	United Kingdom	Europe	0.72	II
915	Llangeneth, Wales	United Kingdom	Europe	0.71	II
916	St Govans, Wales	United Kingdom	Europe	0.69	II
917	Whitesands, Wales	United Kingdom	Europe	0.68	II
918	Newgale, Wales	United Kingdom	Europe	0.66	II
919	Tenby South, Wales	United Kingdom	Europe	0.57	III
920	Southerndown, Wales	United Kingdom	Europe	0.54	III
921	Amroth, Wales	United Kingdom	Europe	0.54	III
922	FreshWater West, Wales	United Kingdom	Europe	0.46	III
923	Blue Lagoon, Wales	United Kingdom	Europe	0.45	III
924	Wisemans Bridge, Wales	United Kingdom	Europe	0.34	IV
925	Broadhaven, Wales	United Kingdom	Europe	0.34	IV
926	Angle, Wales	United Kingdom	Europe	0.33	IV
927	Tenby North, Wales	United Kingdom	Europe	0.26	IV
928	Aberdovey, Wales	United Kingdom	Europe	0.24	IV
929	Towyn, Wales	United Kingdom	Europe	0.23	IV
930	Pothcawl, Wales	United Kingdom	Europe	0.20	IV
931	Saundersfoot, Wales	United Kingdom	Europe	0.15	IV
932	Llantwit, Wales	United Kingdom	Europe	0.04	IV
933	Ogmore, Wales	United Kingdom	Europe	0.03	IV
934	Gileston, Wales	United Kingdom	Europe	−0.20	V
935	Fairbourne, Wales	United Kingdom	Europe	−0.29	V
936	Bigbury on Sea, England	United Kingdom	Europe	0.65	II
937	North Hallsands, England	United Kingdom	Europe	0.33	IV
938	East Sands Ryde, England	United Kingdom	Europe	0.18	IV
939	Giants Causeway, N Ireland	United Kingdom	Europe	0.85	II
940	Magellan Foreland Tip, N Ireland	United Kingdom	Europe	0.56	III
941	Magellan Foreland, N Ireland	United Kingdom	Europe	0.30	IV
942	Stonehave, Aberdeen, Scotland	United Kingdom	Europe	0.25	IV
943	St Ninean, Shetland, Scotland	United Kingdom	Europe	0.98	I
944	Scousburg, Shetland, Scotland	United Kingdom	Europe	0.40	IV
945	Skaw, Shetland, Scotland	United Kingdom	Europe	0.97	I

(continued)

No	Coastal site name	Country	Continent	D value	Class
946	West Sandwick, Shetland, Scotland	United Kingdom	Europe	0.80	II
947	Sagg Harbour Main Beach	USA	America	1.10	I
948	Haven Beach	USA	America	0.71	II
949	La Jolla	USA	America	0.23	IV
950	Alki Beach	USA	America	0.20	IV
951	Cua Dai Beach	Vietnam	Asia	0.22	IV
952	Thuan An Beach	Vietnam	Asia	0.34	IV

References

Anfuso G, Martinez del Pozo JA, Nachite D (2010) Coastal vulnerability in the Mediterranean sector between Fnideq and M'diq (North of Morocco). Comptes rendus de l'Academie Bulgare des. Sciences 63(4):561–570

Anfuso G, Williams AT, Cabrera Hernández JA, Pranzini E (2014) Coastal scenic assessment and tourism management in western Cuba. Tour Manag 42:307–320

Anfuso G, Williams AT, Casas Martínez G, Botero CM, Cabrera Hernández JA, Pranzini E (2017) Evaluation of the scenic value of 100 beaches in Cuba: implications for coastal tourism management. Ocean Coast Manag 142:173–185

Avella F, Burgos B, Osorio A, Parra E, Vilardy S, Botero C, Ramos A, Mendoza J, Sierra P, López A, Alonso D, Reyna J, Mojica D (2009) Gestión del litoral en Colombia. Reto de un país de tres costas. In: Manejo Costero Integrado y Política Pública en Iberoamérica: Un Diagnóstico. Necesidad de cambio. In: Arenas P, Chica A (eds) Red Iberoamericana en Manejo Costero Integrado. CYTED, Cadiz, pp 175–209

Boumeaza T, Sbai A, Salmon M, Ozer A (2010) Impacts écologiques des aménagements touristiques sur le littoral de Saïdia, Maroc oriental, Méditerranée. Rivages méditerranéens 115:93–102

Cerviño J, Cubillo JM (2005) Hotel and tourism development in Cuba: opportunities, management challenges and future trends. Cornell Hotel Restaurant Adm Q 46(2):223–246

Cristiano S, Rockett G, Portz L, Anfuso G, Gruber N, Williams AT (2016) Evaluation of coastal scenery in urban beaches: Torres, Rio Grande do Sul, Brazil. JICZM 16(1):71–78

Ergin A (2009) Case study; a holistic approach to beach management at Çıralı, Turkey: a model of conservation, integrated management and sustainable development. In: Williams AT, Micallef A (eds) Beach management: principles and practice. Earthscan, London, pp 355–358

Ergin A, Karaesmen E, Williams AT, Micallef A (2004) A new methodology for evaluating coastal scenery: fuzzy logic systems. Area 36:367–386

Ergin A, Williams AT, Micallef A (2006) Coastal scenery: appreciation and evaluation. J Coast Res 22(2):958–964

Iglesias B, Anfuso A, Utega A, Arenas P, Williams AT (2017) Scenic value of the Basque Country and Catalonia coasts (Spain): impacts of tourist occupation. J Coast Conserv. https://doi.org/10.1007/s11852-017-0570-0

Langley RA (2006) Coastal scenic assessment of the North Canterbury coast, New Zealand. Unpub. MA thesis. Univ. of Canterbury, Christchurch, New Zealand, 65 pp

Manno G, Anfuso G, Messina E, Williams AT, Suffo M, Liguori V (2016) Decadal evolution of coastline armouring along the Mediterranean Andalusia littoral (South of Spain). Ocean Coast Manag 124:84–99

Nachite D (2009) Diagnostic environnemental – PAC Rif central du Maroc, activité GIZC, PAP/CAR, January 2009, 74pp. https://www.academia.edu/6233802/PAC-Morocco-diagnostic_environnemental

Oliveira T, Scherer M, Anfuso G, Almeida F, Diederichsen S, Williams AT (2016) Classification of Coastal Beach scenarios in Santa Catarina Island, Florianópolis – Brazil. Desenvolvimento e Meio Ambiente 39:217–229

Phillips MR, Edwards A, Williams AT (2010) An incremental scenic assessment of the Glamorgan Heritage Coast, UK. Geogr J 176(4):291–303

Pranzini E, Anfuso G, Botero C, Cabrera A, Apin Campos Y, Casas Martinez G, Williams AT (2016) Sand colour at Cuba and its influence on beach nourishment and management. Ocean Coast Manag 126:51–60

Rangel-Buitrago N, Anfuso G, Correa I, Ergin A, Williams AT (2013) Assessing and managing scenery of the Caribbean Coast of Colombia. Tour Manag 35:41–58

Rangel-Buitrago N, Anfuso G, Williams AT (2015) Coastal erosion along the Caribbean coast of Colombia: magnitudes, causes and management. Ocean Coast Manag 114:129–144

Rangel-Buitrago N, Williams AT, Anfuso G, Arias M, Gracia CA (2017) Magnitudes, sources, and management of beach litter along the Atlantico department coastline, Caribbean coast of Colombia. Ocean Coast Manag 138:142–157

Rangel-Buitrago N, Williams AT, Anfuso G (2018) Hard protection structures as a principal coastal erosion management strategy along the Caribbean coast of Colombia. A chronicle of pitfalls. Ocean Coast Manag 156:58–75. https://doi.org/10.1016/j.ocecoaman.2017.04.006

Ullah Z, Johnson D, Micallef A, Williams AT (2010) From the Mediterranean to Pakistan and back – coastal scenic assessment for tourism development in Pakistan. J Coast Conserv Manag 14(4):285–293

UNEP/GPA (2003) Diagnosis of the Erosion Processes in the Caribbean Sandy Beaches

Williams AT (1987) Coastal conservation policy development in England and Wales with special reference to the heritage coast concept. J Coast Res 31(1):99–106

Williams AT (1992) The quiet conservators: heritage coasts of England and Wales. Ocean Coast Manag 17(2):151–168

Williams AT, Khattabi A (2015) Beach scenery, Nador, Morocco. J Coast Conserv Manag 19(5):743–755

Williams AT, Micallef A (eds) (2009) Beach management: principles and practices. Earthscan, London 480 pp

Williams AT, Micallef A, Anfuso G, Gallego-Fernandez JB (2012) Andalucia, Spain: an assessment of coastal scenery. J Landsc Res 37(3):327–350

Williams AT, Rangel-Buitrago N, Anfuso G, Cervantes O, Botero C (2016) Litter impacts on scenery and tourism on the Colombian North Caribbean coast. Tour Manag 55:209–224

Chapter 7
The Management of Coastal Landscapes

Anton Micallef and Nelson Rangel-Buitrago

Abstract Was carried out a review of several management approaches suitable for coastal landscapes. Various reasons why the coastal landscape should be managed are presented, including the tremendous anthropogenic influence, the importance of quality landscape to the public, increasing knowledge on landscape leading to enhanced possibilities for more effective management, the growing availability of landscape-related data and the importance of wise management.

The aims of managing the coastal landscape are explored. Of particular importance, the need for preservation of scenic integrity and its cultural authenticity was identified. Consideration of different aspects of landscape allowed identification of various approaches to management of the coastal landscape, be it via direct management actions/mechanisms or the indirect but important influence of international – local planning policies and designation. These were explored via consideration of:

- The opportunities presented by various landscape-related conventions and policiy strategies (e.g. The UNESCO World Heritage Convention and the European landscape Convention).
- The contrasting facets of coastal landscape constituent elements (landform, land-use and land-cover).
- The Protected Area Management Approach, such as, that presented by the International Union for Conservation of Nature (IUCN).
- The contribution to management by education, training and increased public participation.
- The application of tools evaluating coastal scenery, such as the Coastal Scenic Evaluation System (CSES) described in this book.

A. Micallef (✉)
Euro-Mediterranean Centre on Insular Coastal Dynamics, Institute of Earth Systems, University of Malta, Msida, Malta
e-mail: anton.micallef@um.edu.mt

N. Rangel-Buitrago
Departamentos de Física y Biologia, Facultad de Ciencias Básicas, Universidad del Atlántico, Barranquilla, Atlántico, Colombia
e-mail: nelsonrangel@mail.uniatlantico.edu.co

© Springer International Publishing AG, part of Springer Nature 2019   211
N. Rangel-Buitrago (ed.), *Coastal Scenery*, Coastal Research Library 26,
https://doi.org/10.1007/978-3-319-78878-4_7

7.1 Introduction

Hull and Revell (1989) identified scenery as a subset of a landscape, with the quality of the latter being described by an extensive range of environmental/ecological, socio-cultural and psychological factors (James Hutton Institute 2014). Furthermore, Amir and Gidalizon (1990) suggested that the visual impact on landscape quality is determined by any physical changes introduced to a site and in terms of a scene observed from one vantage point. Throughout this chapter therefore, the term coastal landscape is used inclusively of seascape and coastal scenery (Fig. 7.1); the latter are not used separately except in particular circumstances that were felt to require it.

A question that must be addressed is, **why one needs to manage the coastal landscape?** Though this landscape has changed in landform and land cover as a consequence of natural geomorphologic processes and climatic influences, recent history has ushered in an increasingly strong anthropogenic influence on coastal landscapes. This has occurred not only via land use practices but also by exacerbation of previously mentioned natural changes. For example, natural coastal erosion processes have been enhanced due to construction of sediment-interrupting coastal structures (Pranzini and Williams 2013; Masselink et al. 2014) and climate change has influenced sea level rise and extreme meteo-marine events (Letcher 2015). This has resulted in ever growing alterations to the coastal landscape at a global-scale. Particularly as a consequence of climate change (but also as a consequence of technological progress and socio-economic stressors), faster rates of changes can be expected in the landscape (UNFCCC 2007); this raises the priority of pro-active management.

Another reason why coastal landscape needs to be managed is because it is so often taken for granted (Natural England 2009a). The coastal landscape is something inhabitants have grown up with, something that was always there and that was

Fig. 7.1 A coastal landscape on the central Mediterranean Island of Gozo (Malta)

considered as unlikely to change (much) in a person's lifetime. At the end of the day, a high quality coastal landscape is appreciated and used by many people, irrespective of age, or country, of origin. In this respect, and because the public all relate to the concept of a common right to the coast, this landscape is a common responsibility to influence, not least by a shared vision of how it should be managed. The latter should reflect individual ideals concerning local distinctiveness, sense of place and quality, which is strongly influenced by a growing expectation from present-day society for tourism, recreation and environment (Williams and Micallef 2009). In this context, coastal landscape management should also serve to recognize and implement public aspirations, but within a secure recognition of the importance of not impeding natural processes wherever possible.

In addition, there is today more knowledge regarding landscape composition and to related environmental functions that support human life and economic activity e.g. Beder (2002), Rösch et al. (2013). The coastal landscape has much to offer in this regard, such as, production (e.g. genetic), carrier (e.g. recreation), regulation (e.g. nature protection), and information (e.g. scientific education) functions (van der Maarel 1978, 1979; de Groot 1992; Cendrero and Fisher 1997; Micallef and Williams 2003).

Given that mankind is more knowledgeable with respect to the importance of landscapes for its wellbeing, coastal landscape management can be seen as an important component of a sustainable way of life. The increased availability of visual landscape quality-related data via an array of techniques including, in addition to the techniques described in Chaps. 3 and 4, remote sensing and GIS modelling (Ayad 2005; Ward and Snoberger 2009; Saeidi et al. 2017) has encouraged land use planners to acknowledge the significance and potential contribution of aesthetic value and visual dimension-related information to the planning process (Ayad 2005). It follows that the provision of scientifically sound assessments of landscape scenic characteristics may additionally contribute to landscape management.

Wise management will also serve to prioritise rare and threatened coastal landscapes and to recognize their cultural and natural heritage value (Burger 2000). This will also serve to address the issue of reduced ecological integrity represented by the resultant mix of natural and degraded components of many landscapes as a result of ever increasing developmental pressures. It will also serve to enhance the potential for conserving biodiversity in landscapes of reduced ecological integrity.

7.2 The Aims of Coastal Landscape Management

Busquets and Cortina Ramos (2017) identified the main aims of landscape management (Table 7.1); these are equally applicable to the coastal landscape.

In a comprehensive presentation of protected area management guidelines for the International Union for Conservation of Nature (IUCN) Category V landscapes (Protected Landscape/Seascape), Phillips (2002) made specific reference to the conservation of scenic values, clarifying that such management should work towards

Table 7.1 The main aims of landscape management

MAIN AIMS OF LANDSCAPE MANAGEMENT
Protecting existing landscapes and creating new ones exhibiting desirable/rare attributes.
Using landscape values and prospects to promote growth.
Improving living standards via development that respects landscape.
Attaining more effective local spatial planning.
Identifying procedures for integrating landscaping with urban design and sectorial policies.
Achieving landscape quality objectives via elaboration of applicable standards, methodologies and tools.
Exploiting the economic, natural and cultural heritage value of landscape in individual areas.
Instituting consensus by initiating social discussion on landscape.
Aid choice and strategy formulation via discussion and arbitration on landscape.

Adapted from: (Busquets and Cortina Ramos 2017)

preserving scenic integrity and maintaining cultural authenticity. The recently updated operational guidelines for implementation of the World Heritage Convention (UNESCO 2016), reaffirms this concept. From these guidelines, one may conjecture that the authentic nature of the scenic properties of a coastal landscape is maintained (via appropriate management) when they are *"truthfully and credibly expressed through a variety of attributes"*, which according to UNESCO (2016, 18) include:

- Form and design.
- Materials and substance.
- Use and function.
- Techniques and management systems according traditions.
- Location and setting.
- Spirit and feeling.

In the same vein, it may be inferred that the integrity of scenic properties may be considered as a function of the wholeness and intactness of the coastal landscape and its attributes. The integrity of a coastal landscape may be examined by assessing the level to which it:

- Contains all elements necessary to express its significance.
- Is of a size sufficient to represent the features and processes which describe its importance.
- Is subject to negative/positive effects of development and/or neglect.

These considerations of scenic integrity and authenticity may be used to counter potential adverse impacts by unavoidable change occurring in protected (or otherwise) coastal landscape. Phillips (2002) suggested that development-related policies addressing protection of scenic values may therefore be either of the type that would compromise scenic integrity and/or authenticity or strengthen them and that furthermore, scenery could be best safeguarded via other lines of policy, such as, those related to consumers of resources that readily influence scenic quality, land use control and protection of natural and historic assets.

7.3   Approaches to Coastal Landscape Management

Management of the coastal landscape may be achieved by various approaches, each describing different aspects of landscape considerations. It may also be protected by direct management approaches and actions as well as the indirect influence of ancillary mechanisms such as national and local planning policies and designation. In the United Kingdom (U.K.) for example, the following designator mechanisms are used (Dorset and Devon County Council 2009).

- Areas of Outstanding Natural Beauty (AONB), such as, the famous north Cornish coast headlands at Tintagel (Fig. 7.2) and Pentire, that are recognized under the National Parks and Access to the Countryside Act, 1949, and granted legal protection due to their landscape quality.
- Sites of Special Scientific Interest (SSSI), are acknowledged under the Wildlife and Countryside Act, 1981, and granted legal protection as a consequence of their importance for wildlife and/or earth science. Such a site includes the Blue Anchor to Lilstock Coast in West Somerset, and the Ballymacormick Point in Northern Ireland.
- Geological Conservation Review sites (GCR), identified via the 1977–1990 Earth Science sites of national and international importance in Britain initiative and considered to embody the most important geological and geomorphological sites within the UK. Several coastal locations on the Western isles of Scotland such as Loch Bee Machair, provide excellent examples of such GCR sites.

Fig. 7.2   Tintagel coast, Cornwall, U.K

In this context and describing coastal cliff conservation and management of the Dorset and East Devon Coast World Heritage Site, May (2015) concluded that legal protection offers the more important tool for management and protection of this site (see Chap. 2). It is a prime player in the conservation of this coastal landscape that most visitors visualize in terms of its species, habitats and ecosystems. However, Barton's (1998, 32) comment remains pertinent today. *"There is a growing conflict between geological conservation bodies seeking to retain the natural state of actively eroding cliffs and local authorities (encouraged by their residents) wishing to prevent marine erosion and cliff-top recession"*. People living inland and those living on the coast have opposing management interests. The former ask for landslide and flood control, the latter obtain benefits from these natural processes that produce socio-economically desirable beach nourishing sediments. Resolving such opposing stakeholder interests is key to a holistic and acceptable approach to landscape management that is part of Integrated Coastal Zone Management. Estrada-Carmona et al. (2014), suggested that coastal landscape management may be achieved through low impact development, zoning and environmental protection regulations and various land use control mechanisms, such as, land purchase, establishment of rights of use and transferable development rights, the latter referring to a zoning technique that offers protection by redirecting development to areas already planned to accommodate growth and development.

Reflecting on the potentially complex nature of landscape management, Terkenli (2010) suggested that landscape should be managed as:

- A repository of collective meaning and symbolism through time;
- An integral part of our natural and cultural heritage;
- Policy-relevant issues in contemporary urban and regional planning and resource management;
- A complex reality developing in historical time and thus embodying and exhibiting place identity, useful in place promotion;
- A system of energy, material and information flows, interwoven in harmonious environmental wholes;
- Material constructions, reflective of the basic organization of society, culture and economy, ever supporting, representing and affecting ways of life and local knowledge.

The management approaches discussed below are not exhaustive but are considered to present the reader with a general reflection of the topic.

7.3.1 Management of Coastal Landscape by Giving It an Internationally Recognized Legal Voice

A number of international and regional organisations, Conventions and policiy strategies are particularly well-suited for making contributions to the management of coastal landscapes. These would include among others, the 1972 UNESCO

Convention Concerning the Protection of the World Cultural and Natural Heritage (the World Heritage Convention) and the 1992 United Nations Convention on Biological Diversity (the latter refering to the close ties between achieving biodiversity conservation andprotecting cultural landscapes as important habitats for species, communities and ecosystems (Dudley et al. 2006). Similarly, the Pan-European Biological and Landscape Diversity Strategyprovides guidelines for landscape management (Conrad 2010) and the Man and the Biosphere Programme (MAB) promotes interdisciplinary management of fluctuations and exchanges between social and ecological systems within biosphere reserves, that includes coastal ecosystems (Matar and Anthony 2017; UNESCO 2017a).

The European Union's Natura 2000 network works closely with landowners and stakeholder groups to promote the sustainable management of some of Europe's most valuable and threatened species and natural habitats (E.C. 2016) and Ramsar sites designation (e.g. Fig. 7.3, wetlands of international importance designated under the Ramsar Convention) similarly provide an effective mechanism for coastal landscape management. As defined by the Convention, such sites include marshes, peat lands, floodplains, rivers and lakes and coastal areas, such as, salt marshes, mangroves, seagrass beds, coral reefs and other marine areas no deeper than six meters at low tide, as well as human-made wetlands (Ramsar Convention Secretariat 2010). The management approach taken by the RAMSAR Convention includes among other, the integration of wetland site management within broad-scale environmental management planning, including river basin and coastal zone management.

The establishment of geoparks is another mechanism that can contribute to the management of coastal landscapes. A geopark is *"an area of outstanding interest for its rocks and landforms, and where greater appreciation and understanding of*

Fig. 7.3 The Cienaga Grande de Santa Marta Ramsar site, Caribbean of Colombia

that geological heritage can benefit local people and businesses through tourism and education initiatives" (Scottish Natural Heritage 2016, 1). In 2015, UNESCO provided geoparks listed within its new 'UNESCO Global Geopark' accreditation system, with an increased level of recognition, raising it to that of World Heritage Sites and Biosphere Reserves (UNESCO 2017b). UNESCO describes its Global Geoparks networks as being compose of *"single, unified geographical areas where sites and landscapes of international geological significance are managed with a holistic concept of protection, education and sustainable development"* (UNESCO 2017b, 1). A prime example of a coastal geopark is that of the Azores, composed of 121 sites of geographic heritage and marine areas within the nine volcanic islands of the archipelago.

7.3.1.1   The UNESCO World Heritage Convention (WHC)

In 2014 there were 46 (of 981 or 4.7%) marine World Heritage Sites (WHS, Abdulla et al. 2014), and 25 other sites with significant marine/coastal areas or coastal features (Spalding 2012). Considering the significant experience gained and employed by the WHC in the protection of heritage through recognition of established land management systems, laws and local knowledge (Rössler 2000; Conrad 2010), the World Heritage Convention may thus be considered as a significant tool for the protection of many unique marine and coastal ecosystems. The latter can no doubt contribute to the sustainable development of such coastal sites, particularly that of tourism and by so doing, provide jobs, food security and income for local coastal societies.

In December 1992 the World Heritage Committee approved three types of cultural landscapes to be incorporated into the convention, making the World Heritage Convention the first international legal instrument to accept and safeguard cultural landscapes (Rössler 2000). The coastal wetland landscape is often considered as a cultural example, echoing a long-standing and close association between humans and this natural environment. Repeated generations have strengthened their settlements in and around wetlands, often establishing them into cultural landscapes. Today, these coastal landscapes are living chronicles of human settlement and represent an invaluable cultural and historical heritage (Dodouras and Sorotou 2011). Some coastal hunter–gatherer settlements (Brewster et al. 2003) and coastal terraced vineyards (Agnoletti et al. 2013) similarly represent such a coastal cultural landscape (Fig. 7.4).

The cultural landscapes added to the Convention in 1992 comprised (Rössler 2000):

- The landscape designed and created intentionally by humans.
- The organically evolved landscape (reflecting the progression of evolution in their structure and component features; this included the relict/fossil landscape and the continuing landscape, where the evolutionary course of development is still in progress).
- The associative cultural landscape indicative of powerful religious, artistic or cultural associations.

Fig. 7.4 A coastal cultural landscape represented by terraced vineyards in the Cinque Terre, Liguria, Italy, with the village of Manarola in the background (Photo Credit: Pierluigi Brandolini)

A handbook for the conservation and management of World Heritage Cultural Landscapes (Mitchell et al. 2009) presents a management framework that reflects inter-related components of guiding principles, management processes and sustaining management via effective governance, funding strategies and capacity building. In this handbook, management of cultural landscapes was presented as one steering change in such landscapes while preserving important facets for present and future generations.

A set of three principles set to guide planning and management actions included (Rössler 2004):.

- People associated with the cultural landscape are the primary stakeholders for stewardship.
- Successful management is inclusive and transparent, and governance is shaped through dialogue and agreement among key stakeholders.
- The value of the cultural landscape is based on the interaction between people and their environment; and the focus of management is on this relationship.

In its operational guidelines for the implementation of World Heritage Convention, UNESCO (2016) describes an effective management plan as one that considers both the natural and cultural aspects of the site and involving formal and informal traditional practices and urban and regional planning instruments and impact assessments for any potential interventions. The UNESCO (2016) management guidelines cite as chief components of such a management plan:

- The use of participatory planning and stakeholder consultation to achieve a comprehensive understanding of the site shared by all stakeholders.
- A typical integrated management sequence of planning, implementation, monitoring, evaluation and feedback.

- An evaluation of site vulnerability to modifications and pressures and monitoring of impacts of emergent trends and suggested actions.
- Tools to ensure a coordinated partner/stakeholder participation and actions.
- Provision of required resources.
- Capacity building.
- A clear and accountable management system.
- A risk preparedness strategy.

The WHC potential for site protection is further strengthened by its ability to delist any site where the entity responsible for its management (State Parties) fails to adequately and timely react to human action(s) that will reduce its intrinsic qualities. Examples of coastal World Heritage sites that have benefitted from different aspects of this type of management include: the 2001 designated Dorset and East Devon coast in England that includes part of the broader and spectacular Jurassic Coast on the English Channel coast of southern England; the quintessential Mediterranean landscape of the Amalfi coast (Costiera Amalfitana, UNESCO 2017c) on the mid-south-western coast of Italy, inscribed in 1997 and representing a combination of natural and cultural coastal aspects (Fig. 7.5); the Ningaloo Coast on the north-west Australian coast with its distinctive 260 km reef and the Ecuadorian Galapagos islands (Archipelago de Colón) in the Pacific Ocean (UNESCO World Heritage Centre 2017).

7.3.1.2   The European Landscape Convention, 2000

More specifically, the European Landscape Convention (ELC) has been designed for the protection of landscapes (Kuleshova and Semenova 2012) and provides an optimal mechanism for management of the coastal landscape. Specifically, Article 3 of this Convention (that came into force in 2004) aims to: *"promote landscape protection, management and planning, and to organise European co-operation on landscape issues"* (Council of Europe 2000, 3). This international treaty is the first to be wholly concerned with European landscapes, and encourages E.U. Member States to develop a national landscape policy not limited to the protection of exceptional landscapes but also those ordinary, particularly areas most affected by economic, social or environmental pressures, including urban, peri-urban, industrial and coastal areas (Council of Europe 2012). The Council of Europe also emphasises that the ELC considerations on landscape are underlined by the Council's stress on human rights and democracy; in this regard, the convention highlights that while it promotes societal rights to landscape quality, this comes in tandem with societal responsibilities towards achieving this goal, including that of landscape management.

The ELC definition of landscape management is *'an action, from a perspective of sustainable development, to ensure the regular upkeep of a landscape, so as to guide and harmonise changes which are brought about by social, economic and environmental processes'* (Busquets and Cortina Ramos 2017, 25). In their propos-

Fig. 7.5 The Amalfi coast, Italy (Photo Credit: Aldo Cinque)

als for implementation of the ELC, the authors presented landscape management as a strategy formulation process, enabling landscape and quality of life improvement within a sustainable development and 'landscape management project' framework. In this context, landscape management was presented as reflecting four dimensions important to the ELC, namely:

- The social dimension of landscape management that is considered a reflection of the societal interaction with their natural environment; it requires recognition of public perceptions and preferences concerning landscape and civic participation in all aspects of the management process.
- The sustainable dimension concerns the inclusion of sustainable development principles.
- The operational dimension reflects the need for an action oriented instrumental type of management.
- The temporal nature of landscape management is in recognition of natural and anthropogenic induced landscape change.

In its efforts to establish an institutional framework that would encourage a landscape viewpoint to influence spatial planning, land use and resource management, Natural England (an executive non-departmental public body in England that is sponsored by the Department for Environment, Food & Rural Affairs to advise the government on the natural environment) developed guidelines for implementing the ELC (Natural England 2009b). Table 7.2 presents a checklist of the ELC principles that were developed to be used together with the guidelines for the all-important integration of the intent of the ELC into plans, policies and strategies.

Adopting an interestingly holistic and integrated perspective, Schaaf and Clamote Rodrigues (2016) have proposed an exhaustive and extensive list of recommendations to harmonize the management of multi-internationally designated areas (MIDAs) including Ramsar Sites, World Heritage Sites, Biosphere Reserves and UNESCO Global Geoparks. These recommendations are specifically aimed at (local) site managers, (national) designating instrument focal points and stakeholders and (international) designating instrument decision-making bodies and secretariats. The authors outlined several potential advantages, as well as management challenges of having areas with more than one designation. Among the benefits, they refer to enhancement of resilience to peripheral stresses, involvement of local stakeholders, highlighting scientific value towards research, education, public awareness raising and trans boundary cooperation, consolidation of inter-institutional cooperation, fundraising and increasing popularity and standing and consequential local economic development. However, several challenges to the management of MIDAs were also identified, including the need to address the potential of:

- Non-coherent legal and administrative structures and coordinated policy and intervention instruments for multiple national authorities working on the same site
- Competition for national and international funding.
- Dissimilar reporting requirements by the various designating organizations and potentially complex information exchange between the latter, management agencies and national authorities.
- Different designating mechanism objectives, strategies and compatibility with site geographic delineation.

Table 7.2 Checklist of European landscape convention principles to aid implementation into plans, policies and strategies (Natural England 2009b)

Checklist for implementing the European landscape convention into plans, policies and strategies, policies and strategies
1. Be clear in the use of landscape terms and definitions
The plan, policy or strategy should:
Use/relate to definitions set out in the ELC
Use the term landscape explicitly rather than other terms such as 'countryside', 'rural', 'natural environment' etc.
2. Recognise landscape in a holistic sense
The plan, policy or strategy should:
Recognise landscape in its own right
Recognise landscape as a whole involving the interaction of natural, cultural and perceptual factors
Recognise that landscape exists at all scales
3. Apply to all landscapes
The plan, policy or strategy should:
Apply to the entire area or place covered by the plan
Apply to all landscape – outstanding and ordinary
4. Understand the landscape baseline
The plan, policy or strategy should:
Draw on the appropriate hierarchy and level of landscape knowledge (evidence).
5. Involve people
The plan, policy or strategy should:
Use appropriate techniques to involve people in:
Identification and assessment of landscapes.
Understanding what is valued about any given landscape.
Establishing objectives for the landscape.
Establishing policies – for protection, management and planning.
Monitoring change.
A decision will need to be made at the outset of the process about who to involve and when.
6. Integrate landscape
The plan, policy or strategy should:
Promote multifunctional landscapes
Integrate landscape into all sectorial policies that have a direct or indirect influence on landscape
Consider any defined landscape objectives for any given geographic area.

- Local apprehension to land-use options and confusion by residents and visitors to designation significance and status as a consequence of (multi) international designation.
- Erosion of conservation potential by unsustainable tourism use as a consequence of increased popularity.

Lino and Britto de Moraes (2004), in their study on the protection of landscapes and seascapes in coastal regions of Brazil, considered two protected sites having distinctly different designations, the Mata Atlantica Biosphere Reserve and the federally designated (as an Environmental Protection Area) Cananeia-Iguape-Peruibe coastal area. In their conclusions, the authors identified not only the important role of such designations in the protection of local coastal landscapes and seascapes but also the complementarity of such differing designatory mechanisms, in the sustainable management of such coastal areas (including local stakeholder involvement and a managed urban growth). In a similar study on protected landscapes and seascapes but considering in particular their relevance to Small Island Developing States (SIDS), Romulus (2004) identified that in its exploration for a mode of sustainable development that works within its socio-economic constraints, the Caribbean has identified that opportunity by turning to an efficient management of its natural and cultural heritage, within the context of the 'St. George's Declaration of Principles for Environmental Sustainability', that states among other ... *the effective management of environmental resources at local, national and international levels is an essential component of sustainable social and economic development, including the creation of jobs, a stable society, a buoyant economy and the sustaining of viable natural systems on which all life depends* (OECS 2000, 1).

The management of coastal landscapes has also benefitted from government-supported initiatives. For example, the Scottish Natural Heritage (the U.K. government's adviser on all aspects of nature and landscape across Scotland) has developed guidelines to assist professionals assess possible landscape and visual impacts of development on Scotland's coastal natural and cultural heritage (Land Use Consultants 2017). The utility of such guidelines were identified as including consideration of coastal assets and characteristics (e.g. maritime influences and the type of the coastal edge) within new spatial and development plans, new projects, aquaculture and overall marine planning.

7.3.2   The Management of the Coastal Landscape via Consideration of its Constituent Elements

Landscape has been described as being composed of three main elements, namely the natural elements of landforms and water bodies, the living element of land cover and the human element of land use (Anderson 1979; Education Services Australia 2013). Aspects of this approach have already been hinted at in Chap. 3 in the form of Linton's (1968) composite assessment of scenery based on landform landscape and land usage – the building block of Chap. 4; the Coventry-Solihull-Warwickshire, UK (1971) index of landscape value derived from the assessment of landform, usage and land feature measurements and consideration of landform and land use area parameters by the Zube et al. (1974) scenic evaluation study. To this end, the management of the coastal landscape would include the management of coastal landforms and water bodies, of coastal land uses and of vegetation land cover.

The evaluation of coastal scenery therefore is a vast subject that embraces disciplines, such as, geology, geography, botany, ecology, sociology, etc. and it *'depends greatly on that of the country to which it forms the margin'* (Steers 1972, xi). Physiography is one of the most important constituent elements and exhibits great variation, e.g. in the UK, coastal cliffs vary from those composed from a mountainous igneous rock hinterland – Waterstein Head, Skye; or hard Carboniferous limestone – Great Orme's Head, Wales, in stark contrast to an area composed of soft glacial material – Sheringham cliffs, Norfolk. Spectacular cliffs at Canna island near Skye, Scotland, UK are formed from bedded tuffs and agglomerates alternating with sheets of columnar dolerite lavas. If extensive folding has taken place, the rock trend can parallel to the coast – central Cardigan Bay, UK, but at right angles in western Pembrokeshire. Therefore *'the shoreline is the outcome not of one isolated event but of a whole series'* (Steers 1972, 254). In sedimentary areas, structure e.g. dip, strike and bedding are very important in determining the resultant scenery apart from areas where blown sand and drift deposits are in the majority. Sand/shingle areas, salt marshes, anthropogenic features ranging from urban conglomerations, to remote areas, along with the rise of telecommunication towers, unsightly/sensitive developments, etc. all add/subtract to an expression of scenery that one perceives.

7.3.2.1 Management of Coastal Landforms and Associated Geomorphological Processes

Composed of a wide range of erosional and depositional landforms, management of the coastal landscape necessarily involves management of coastal landforms and associated processes. Depositional landforms forming sedimentary coastal landscape include among other beaches, dunes, spits, tombolos, barrier islands, sandy river mouths, and mudflats. Erosional landforms on the other hand are largely representative of the rocky coastal landscape and include a range of elements (Fig. 7.6), such as, headlands, sea cliffs stacks and caves, rock arches and shore platforms (Masselink et al. 2014). An additional coastal landscape is that represented by offshore islands. The contribution of coastal landforms to the enhancement of coastal scenery is fully recognized by the Coastal Scenic Evaluation System (CSES) (described in Chap. 4). Coastal scenic rating by this technique is influenced by both the presence of natural coastal landforms but also by the richness of some of these features; for example, a cliff is assessed not only by its height, but also by its slope and special features, such as, indentation, banding, folding, scree, irregular profile, etc.; a beach by its type, width and colour and a rock shore, by its slope, extent and roughness (Chap. 4, Table 4.1).

Management of such coastal landforms may be considered in its simplest sense, as that cognizant of the action of a variety of fluvial, aeolian and coastal processes in the generation, transport and deposition of coastal sediments i.e. coastal landform management may be considered as being composed of either hard or soft engineering or do nothing approaches (Pilkey et al. 2016; Williams et al. 2018). In practice, however, management of the coastal landform (often one specifically addressing coastal erosion and/or flooding) is one determined by policy (Doody et al. 2004). The latter further suggested that a number of socio-economic, political and environmental factors

Fig. 7.6 Spectacular rocky coastal landscape on the north-west coast of the central Mediterranean island of Gozo (Malta), including headland, cliffs, stack, caves, and rock arch

will influence the type of management strategy chosen; these include threatened dwellings and or coastal resorts representing large financial investments, government infrastructure such as energy generating installations, degree of sea-level change, local geomorphology and sediment budgets. In this context, Micallef et al. (2017) suggested that it is natural to identify many instances where countries e.g. have adopted a mix of such management policies and that choice was ultimately a strategic one highly influenced by considerations of financial and sediment availability.

Despite some on-going misgivings regarding its elusive nature in practice (Smith et al. 2011) the use of an Integrated Coastal Zone Management (ICZM) framework to manage coastal areas is now well-documented (European Commission 2012) and legally established by, for example, the Barcelona Convention protocol on ICZM (UNEP/MAP/PAP 2008) and the European Commission's 2000 Recommendation on ICZM and 2013 Directive establishing a framework for Marine Spatial Planning & ICZM (Meiner 2010). Another approach that may be used to effectively manage coastal landforms and related coastal processes involves the use of coupled numerical modelling to answer questions related to application of effective coastal management (Whitehouse et al. 2017).

Another method of particular relevance to management of coastal landform and processes is the development of Shoreline Management Plans (SMPs) that are now widely implemented, for example, along the coast of the United Kingdom (DEFRA 2006). They are based on a large-scale assessment of coastal risks (flooding and erosion) associated with coastal processes. Sediment cells created by features that limit sediment transport via long-shore drift patterns are identified, allowing prediction of the potential influence of human actions on sediment movement (and related erosion) within individual cells. To this end, the generation of SMPs involves consideration of various aspects of coastal geomorphology, geology, ecology, exposure, flood and erosion risk, coastal protection type, and management strategy (Niemeyer et al. 2016). Similarly, in Italy, sediment cells are the basic coastal management unit in the "Linee guida nazionali per la difesa della costa: gestione della dinamica costiera" (Guidelines for shore protection: managing coastal dynamics, ISPRA 2016).

With regard to coastal management policy, a number of options are available to coastal administrators (Doody et al. 2004; Williams et al. 2018); these include:

- **Hold the line** is a policy option referring to the intention of either constructing new or maintaining existing artificial coastal defence works, such as, seawalls, rock armouring or off-shore/submerged breakwaters. Alternatively, soft engineering works such as beach nourishment and/or dune management can also be used in this option.
- **Advance the line** refers to the construction of defence works seaward of existing ones.
- **Managed retreat/realignment**. Normally applied in low-lying coastal areas, this policy option adapts to natural coastal processes by realigning the coast. Engineering solutions establish a new line of defence landward from the current position while allowing a natural but managed shoreline movement that normally involves flooding of low value land. The latter will increase wave energy buffering (protecting the coastal area in question) but may also generate new sediment to protect down-drift areas.
- **Limited intervention** is a policy option leaning to accommodation to natural processes. In this case, coastal interventions adapt vertically by elevating coastal land and buildings to cope with inundation.
- **Do nothing**. This policy opts to provide no protection. Any existing defence works are not maintained further leading to eventual abandonment. In this case however, the long-term resultant ruins and associated rubble has a tremendous negative effect on landscape value. As a consequence, some of the worst coasts from a scenic point of view are those where shore protection structures were left unmaintained (Fig. 7.7).

Hard engineering schemes include the construction of seawalls, the use of gabion elements, emerged and submerged groynes, revetments, rock armouring (rip-rap) structures and off-shore/submerged breakwater, while soft engineering considers beach nourishment, sand-dune stabilization and managed retreat. While the general trend has been towards favoring soft coastal management options due to their reduced interference with coastal sediment transport systems and pathways, the selection of an appropriate coastal protection system is dependent on many factors, including site location, the type of protection desired, general environmental conditions, accessibility to construction materials, expertise in design and construction and accessible funding. Pranzini and Williams (2013), have given an excellent account of coastal erosion and resultant protection measures in Europe. Apart from

Fig. 7.7 A very low coastal scenic quality influenced by poorly maintained or abandoned coastal protection structures (Salgar, Atlantico, Colombia)

managed retreat/realignment and cliff stabilization works that involve matting and vegetation planting to increase cliff face stability, soft engineering schemes involve beach nourishment. This shoreline management approach is based on the understanding that a natural beach represents an extremely efficient system for wave energy diffusion. To this end, large quantities of sand are normally added to either existing but partially eroded (poorly functioning) beaches or on erosion/flooding threatened shorelines with no beach. Unlike the majority of hard structures however, the generated beach is visually appealing and adds to the recreational potential of the coast. On the negative side, economically feasible sources of suitable marine sediment are becoming more difficult to access (Micallef and Cassar 2001) and beach nourishment is potentially expensive due in part to often needed re-nourishment arising from natural beach sediment losses. This may be offset by replacing sand nourishment with granules and/or pebble fractions that are much more resistant to erosion (Cammelli et al. 2006).

Another innovative alternative to individual beach nourishment projects is the 'Sand engine/motor' concept recently developed in the Netherlands. Following a 'building with nature' concept, the Dutch government (Rijkswaterstaat and the provincial authority of Zuid-Holland) created a hook-shaped peninsula on the TerHeijde coast, via deposition of about 21.5 million cubic meters of dredged sand between March and November 2011. This artificial peninsula extended 1 km into the sea and spanned 2 km wide at the near-shore. The expectation was to replace a 20-year period of five-yearly beach nourishments on several beaches with a single operation that caused less disturbance to the sea-bed and allowed natural long-shore sediment distribution to take its course. The first project assessment (5 years after final sand deposition) has indicated very positive results in the form of increasing coastal safety, creating additional space for leisure and nature and contributing to coastal management-related knowledge. Similar pilot schemes are being developed in Jamaica (West Indies) and in Norfolk, on the east coast of the U.K. (De Zandmotor 2017).

7.3.2.2   Management of Coastal Landuse

Coastal land use has a major influence on coastal landscape scenery. It comprises the human element of landscape and its impact may be best described by its influence via deforestation, agriculture (and sometimes associated drainage of wetlands), tourism development, urbanization, river damming and re-routing (resulting in reduced coastal sediments) and physical coastal defense works (Barrachina 2007; Eurostat 2017). The agricultural revolution gave rise to increased population growth, with a resultant larger scale agriculture, increased risk of forest fires, clearing of vegetation, deforestation and urbanization. An increased practice of animal husbandry resulted in overgrazing which, together with deforestation, resulted in increased soil erosion. Human land use influence was however not always negative! Certainly coastal conservation projects and reforestation contributed positively and the establishment of agriculture around hillsides gave rise to typical hill-slope terraced landscape that when present, has a dramatic influence on coastal landscape scenery (e.g. the Liguria coast in Italy).

There are several mechanisms that address management of coastal land use. At the European level for example, both ICZM as well as Maritime Spatial Planning are endorsed as effective instruments for this purpose (VLIZ 2013; European Commission 2016, 2017) by both E.U. related directives and regional conventions, such as, the Barcelona Convention for the Mediterranean Sea and its protocol on ICZM. Spatial Planning is concerned with more than just wise land-use planning, as it also attempts to resolve numerous and often conflicting sectorial e.g. transport, agriculture and environment policy demands (UNECE 2008). Cullingworth and Nadin (2006, 91) have defined spatial planning as being concerned with *"the problem of coordination or integration of the spatial dimension of sectorial policies through a territorially-based strategy"*. It therefore encourages a more balanced distribution of activities in often-limited land and sea space by addressing competing policy demands. Spatial planning is one of the more important tools used for coastal management (Felter and Morris 2016). In identifying the role of spatial planning in ICZM, BaltCoast (2005) suggested that in many cases of coastal development, the distinction of issues spatially, generated problem resolution earlier in the ICZM cycle; spatial planning was also identifies as able to contribute to ICZM success as a consequence of its:

- Ability to reconcile different sectorial needs and provide a multi-agency approach.
- Common (to ICZM) problem identification, planning, implementation and evaluation cyclical framework.
- Suitability as an information platform able to satisfy multi-level (including cross-border) and multi-institutional requirements.

- Ability to offer a spatially categorised GIS-based data organisation system that aids multiple geographic area scale representation and data exchange.
- Track record in stakeholder participation and conflict management.
- Ability to project resource use and conservation conflict resolution over long-term scenarios.

A number of techniques such as landscape ecology that attempts to integrate varying land uses and land users (as in watershed management) have influenced the development of an increasingly popular technique that can be used to address land use management, namely Integrated Landscape Management. This approach involves multiple stakeholders, who work together to integrate policy and practice for their individual land use objectives (Reed et al. 2015; Denier et al. 2015). While the ultimate aim of this approach is that of realizing sustainable landscapes,it does so mainly by realizing landscape multi-functionalityand attempting to restrict potential conflicts and impacts among the various land use functions. In this sense, the coastal landscape can, if so managed,simultaneously and harmoniously serve not only to provide a recreational function, but also one of ecosystem conservation, human livelihood and spiritual or artistic inspiration.

An ICZM framework is the ideal space to develop integration. This framework stimulates complete incorporation of different coastal activities by coordinating government and the private sector (Olsen et al. 1997; European Commission 2012). Combining coastal scenery management usage and ICZM is an excellent opportunity to strengthen several integrative efforts.

Adequate coastal management and conservation requires that the present use of coastal resources must meet the needs of the population, without endangering the ability of future generations to respond to their needs. Any coastal scenery management strategy requires implementation of effective and efficient solutions based on scieintific knowledge and includes coastal users' priorities and preferences identifying, maintaining and, where possible, enhancing the value of the coast to the economic, environmental and social well-being of local communities (Komar and McDougal 1988). A short-term perspective conditioned by economic considerations manifested in an action-reaction basis or a cost-benefit analysis approach does not work under this framework. Clearly coastal scenery management must be focused on identifying the problems that affects the coast together with implementation of strategies from a regional and long-term perspective.

Under current conditions, coastal scenery management plans are required. The integrated coastal management policy cycle formulated by Olsen et al. (1997) is a good starting point to achieve this purpose. With specific respect to scenery this cycle maybe adapted to include:

- Identification and understanding of process that affects the scenery.
- Preparation and planning of strategies.
- Managing and evaluation.

In the last two stages, the following steps can be usefull:

- Consider scenery in vulnerability assessments of coastal communities.

- Develop scenarios and tools that model impacts over scenery.
- Build decision support systems to help communities visualize impacts and possible solutions.
- Identify stakeholders perceptions and priorities with regard scenery.
- Enact policies to ensure scenery integrity of the coast.
- Incorporate proven scenery management interventions.
- Strengthen the capacity for scenic evaluation.

The above can be included in policies supporting strategies and above all, must be constructed with enough robust scientific research for decision makers to make rational decisions. Political support is also required that helps integration of diferent tools into coastal scenery management strategies. An adequate financial, knowledge and technological resources allocated respectively for integrating scenery management in national portfolios is recommended. This must be included in national policy-setting and awareness raising, planning, practices and, capacity building. Several key features of coastal scenery management make this a competitive strategy in a wide range of settings:

Economy High capital investments derived from an expensive infrastructure can be avoided. e.g. the use of ecosystems in scenery management can have lower long-term maintenance costs, and some can take less time to implement. Savings derived from natural infrastructure can be quite significant.

Extra Benefits Implementation of this strategy also improves water quality, supports recreation, provide habitats and delivers economic and other societal benefits.

Multiplicity These projects can be developed successfully in a wide range of habitats and coastal types. Knowledge of coastal systems may guide implementation of the strategy.

High Community Value This strategy can be very familiar with local communities, and can be critical in raising new project funds and building political support.

Flexibility Under the right conditions, coasts have the capacity to adapt by themselves.

Complementary Support Existing restoration and conservation projects can be enhanced.

A considerable level of cooperation is required from all stakeholders to take action on lanscape degradation. This will help foster and narrow existing links between coastal management, disaster risk reduction and climate change adaptation, as well as between science and policy. Coastal management authorities need a robust and clear management framework to resolve issues related to scenery degradation. Rangel-Buitrago et al. (2015, 2018) suggested these need to be focused on the following specific points:

- Determination of significant scenery hot spots.
- Identification of scenery impacts.
- Comprehension of processes involved in scenery degradation.
- Enhance community involvement on scenery issues and related problems.
- Evaluate and determine the optimal options for management taking into account viewpoints of all stakeholders.

This same authors also suggests that Government institutions can achieve such objectives by undertaking a law reform agenda based on five key elements:

- Update current environmental legislation with strong management rules focused on landscape degradation issues, which also must be implemented.
- Add new measures to better support decision-making, including a decision support framework including a scenery management manual.
- Reducing individual consequences over the environment-related to bad management practices.
- Raising public awareness about risks associated with landscape degradation.
- Providing an efficient and sustained response to landscape issues.

Figure 7.8 illustrates the steps to follow in the scheme taking into account aspects under the well-known integrated coastal management policy cycle formulated by Olsen et al. (1997). This plan must be included in supporting strategies, politics and must be built with enough scientific research for decision-makers to make rational decisions. This plan can achieve continual improvement of scenery management.

However, a weak coastal governance, bad political practices, an inadequate financial commitment and the nature of public participation can hinder the formation of ICZM regulations, and have made it a challenge to incorporate coastal scenery management into unstable ICZM regimes. As world coastal population increases, increasing pressure is applied to government at all levels to resolve the coastal degradation problem. Unfortunately, from a governance viewpoint, many management strategies fail due to a weak institutional framework, accompanied by diluted and compromised management regulations.

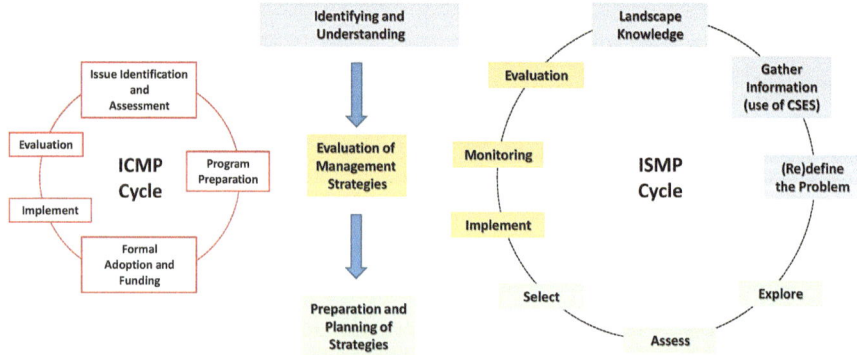

Fig. 7.8 Coastal management cycle (Adapted from Olsen et al. 1997)

Furthermore, the integration of land and marine based spatial planning systems (as a consequence of the land catchment –coastal marine/offshore interdependence) together with a contribution of ICZM (via for example its objectives of achieving participatory planning and stakeholder involvement) and within overarching environmental management concepts, has been suggested as a very effective coastal management approach (Smith et al. 2011). The authors list several important elements that are reflective of the interactive nature of the land-sea interface, such as the influence of changing coastal land uses on associated landforms, habitats and species; the modifications to coastal landscape resulting from coastal development (e.g. marinas) and coastal protection infrastructure (e.g. offshore breakwaters); and the dependence of coastal populations on strategic maritime industries such as fishing and tourism. They further suggested that integration of the land and marine-based spatial planning systems, may be enhanced by ICZM, Strategic Environmental Assessment (SEA) and implementation of a number of E.U. Directives (e.g. Water Framework, Habitats and Bathing Water Directives). Given this close association between terrestrial and marine components and the resultant impact of anthropogenic activities on coastal and marine ecosystem services, an ecosystem-based approach to coastal and marine spatial management has been promoted (Blue Solutions 2017; Ehler and Douvere 2009). The authors describe ecosystem-based management (EBM) as one where human influence is considered central to entire ecosystem functioning and where all coastal uses and users are therefore assessed. In this context, the aim of EBM is to sustain ecosystem services via a sustainable management of all human activities and sectors.

7.3.2.3   Management of Coastal Land Cover

The role of land use and land cover changes (LULCC) in the transformation of traditional landscapes has been clearly established (Brunori et al. 2017); this is particularly the case within the Mediterranean coastal landscape where a number of human induced changes to natural land cover have taken place. Over the last several decades in particular, the Mediterranean region experienced large scale LULCCs as a consequence of intensification of agriculture in some areas and relinquishment of an agro-pastoral way of life in other (often rural) areas followed by population migration to increasingly urbanized coastal areas (Fig. 7.9), loss of forests to fires and deforestation and, the fast and ongoing growth of tourism and related activities (Burke and Thornes 2004; Sluiter and de Jong 2007). Given this trend of LULCCs, (Feranec et al. 2010) identified potential effects as including loss of agricultural and natural lands, landscape fragmentation and a reduction of biodiversity and in this context, recognized the importance of long-term monitoring of such changes and impacts.

Deforestation and agro-pastoralism has seen the historical virtual destruction of much of the Mediterranean's typical climax forest, such as, the broad-leaved oaks (*Quercus ilex, Q. suber, Q. coccifera, Q. pubescens, Q. cerris, Q. pyrenaica, Q. toza, Q. calliprinos, Q. ithaburensis*) and conifers (*Pinushalepensis, P. brutia,*

P. pinea, P. pinaster and Juniperus sp. Scarascia-Mugnozza et al. 2000). Particularly in coastal areas, forest cover has yielded to an anthropogenically influenced garrigue and maquis vegetation cover; near cities and industrial zones, deforestation has resulted in soil erosion, negatively influencing agriculture (Carpaneto et al. 2003; Hughes 2011). Over its historic human occupation, the Mediterranean basin has experienced significant landscape changes arising in part to the replacement of the above-mentioned indigenous evergreen oak, deciduous and conifer forests (Conservation International 2016) by a semi-natural mix of forests, shrublands, woodland stands and widespread reforested lands; the latter are largely composed of non-indigenous species, such as conifer (e.g. cypresses, radiate pine), poplars and eucalypt plantations (La Mela Veca et al. 2016). The introduction of alien species have also replaced indigenous vegetation and associated landscape. In a comparative study of patterns of plant invasions in the Mediterranean Biome. Arianoutsou et al. (2013) provide examples of naturalized alien shrub and tree species, such as, the southern USA succulent shrub *Opuntiaficus-babarica* found in abandoned grazed fields in Stylida, on the northern shore of the Malian Gulf of Greece and the east Asian tree *Alianthusaltissima* in central Greece.

More recent coastal development in the form of urban sprawl has also replaced both natural and agricultural landscape. Changing coastal area characteristics, population densities in European coastal areas are approximately 10% higher than those inland, and in some countries, reaching 50% European Environmental Agency (2006). Alves et al. (2007) refer to instances where the transition rate from a natural to anthropogenic coastal landscape is disturbingly faster than increasing population densities. A case study of Izmir's (Turkey) coastal landscape change between 1963–2005 (Hepcan et al. 2013) identified large losses of agricultural land cover (13.65–

Fig. 7.9 Coastal urbanization on the Mediterranean Andalusian coast of Spain

5.19%) and a much larger growth of initially compact urbanized areas and subsequently urban sprawl by 2005.

As a consequence of such unsustainable coastal population densities, water over-extraction has in some places resulted in the salinization of water tables; the Mediterranean island of Malta is one such location where such salinization has impacted soil properties and in turn the resultant vegetation cover (Spiteri et al. 2015). Referring to the Mediterranean, water over-extraction has also been reported to have resulted in patterns of desertification in southern Mediterranean coastal areas (Mulligan et al. 2016). More disturbingly, the Aston Centre for Europe (2013) have referred to all sub-regions of the Mediterranean as having already undergone differing degrees of desertification with the very likely prospect of increased future exposure. While the highest desertification level reported was for the Middle East and North African (MENA) region (with Morocco, Lebanon, Syria, Israel and Turkey nearing critical levels), Albania on the northern Mediterranean shore was also reported as facing country-wide exposure to desertification. Coastal wetlands have in many places been replaced by an agricultural land cover as a consequence of reclamation via land drainage, with estimates of 50% of the world's wetlands having been converted for use as agricultural, industrial or urban purposes and into oil palm plantation in South-East Asia (Verhoeven and Setter 2010). Prime examples of such land drainage may be seen in the Tuscan region of Italy within the Maremma National Park (Colombini and Chelazzi 2010). With more than 10,000 kms of coastline, many coastal plains and valley floodplains in Italy have required draining or protecting against floods (Alexander 1985).

Promoting sustainable land use management systems, Bajocco et al. (2012) demonstrated the ability to assess conversion and/or modification LULCCs and trends of land degradation on the coastal area of the island of Sardinia (Italy) by developing a Geographic Information System describing multi-temporal land cover and diachronic land sensitivity maps. In their work, the authors presented a management structure useful for developing Mediterranean coastal land use and environmental protection policy by forecasting land sensitivity response to LULCC s scenarios. Importantly, Geist and Lambin (2004) stress that a better understanding of the land-use decision-making process and context is needed to allow improved land management and consequently, a better-quality landscape. Di Gregorio et al. (2011) described that given the important influence on landscape by LULCCs, their long-term monitoring, by for example, application of the FAO Land Cover Classification System to high/medium resolution time-series land cover data, together with a better understanding of human-related causes and responses is critical in land degradation studies. In this regard, the assessment of the quantity and type of land change occurring and decision-making for sustainable land use management is significantly dependent on detailed land cover and land degradation data (Jansen 2004; Jansen and Di Gregorio 2004).

Application of land use and land cover management in River Basin Management (RBM) sets out to achieve integrated water resources management and contribute to overall landscape management (Wisconsin Department of Natural Resources 2014). RBM plans are a requisite of the European Union's Water Framework Directive that a sustainable use of the European water environment that includes amongs others, coastal waters out to one nautical mile. Similarly, the development of land use and land cover classification-systems facilitate landscape management (Anderson et al. 1976; Pal and Mather 2003). In addition, a number of specific online tools and atlases, such as, the Coastal Area Land Cover Analysis Project (CALCAP), Land Cover Atlas produced by the NOAA Office for Coastal Management (NOAA 2017) and the Ohio (USA) Coastal Atlas and Interactive Coastal Atlas Map Viewer (ODNR 2017a, b) have been developed to present large LULCC data-sets into easy to use coastal management and decision-making related information regarding in what manner and in which areas development is likely to be influencing important ecological and coastal recreation areas. By providing ready processed data of coastal area land cover characterization and measurements on local watersheds, such tools increase such data access to non-technical users and non-GIS experts.

7.3.3 Management of Coastal Landscape via the Protected Area Management Approach

The International Union for Conservation of Nature (IUCN) is an observer and consultant at the United Nations (UN) and makes an important contribution to the management of coastal scenery via implementation of numerous international conventions on nature conservation and biodiversity. In particular, it has developed a set of categories for the management of protected areas (I–VI, Table 7.3), which have become accepted at international (e.g. UN) level and assimilated into national legislation of many countries.

Of the six IUCN management categories of protected areas (Table 7.3), Category V (Protected Landscape/Seascape) is of particular relevance to the management of coastal landscapes and ergo scenery. It has been defined as:

Table 7.3 The six IUCN management categories of protected areas (Dudley 2008)

Category	Description
Ia	Strict Nature Reserve: Protected area managed mainly for science.
Ib	Wilderness Area: Protected area managed mainly for wilderness protection.
II	National Park: Protected area managed mainly for ecosystem protection and recreation.
III	Natural Monument: Protected area managed mainly for conservation of specific natural features.
IV	Habitat/Species Management Area: Protected area managed mainly for conservation through management intervention.
V	Protected landscape/seascape: Protected area managed mainly for landscape/seascape conservation and recreation.
VI	Managed Resource Protected Area: Protected area managed mainly for the sustainable use of natural ecosystems.

Table 7.4   General principles for the management of Category V

General principles for the management of Category V (protected landscape/seascape)	
1	Conserving landscape, biodiversity and cultural values are at the heart of the Category V protected area approach.
2	The focus of management should be on the point of interaction between people and nature.
3	People should be seen as stewards of the landscape.
4	Management must be undertaken with and through local people, and mainly for and by, them.
5	Management should be based on co-operative approaches, such as co-management and multi-stakeholder equity.
6	Effective management requires a supportive political and economic environment.
7	Management of Category V protected areas should not only be concerned with protection but also enhancement.
8	When there is an irreconcilable conflict between the objectives of management, priority should be given to retaining the special qualities of the area.
9	Economic activities that do not need to take place within the protected landscape should be located outside it.
10	Management should be business-like and of the highest professional standard
11	Management should be flexible and adaptive.
12	The success of management should be measured in environmental and social terms.

Protected landscape/seascape, Source: Phillips (2002)

> Area of land, with coast and sea as appropriate, where the interaction of people and nature over time has produced an area of distinct character with significant aesthetic, ecological and/or cultural value, and often with high biological diversity. Safeguarding the integrity of this traditional interaction is vital to the protection, maintenance and evolution of such an area. (IUCN 1994)

More recent definitions (e.g. Dudley 2008) have replaced the opening line, *"An area of land, with coastal and sea as appropriate"* with a more straight forward *"A protected area"*. This Category includes protected coastal landscapes that have received the highest level of anthropogenic influence (see also Chap. 2).

Phillips (2002) has provided detailed management guidelines for IUCN Category V (Protected Landscape/Seascape). Management in this regard, is presented as being based on a set of 12 general principles (Table 7.4).

7.3.4   Management of Coastal Landscape via Education, Training and Increasing Public Participation

Experts and technical and non-technical personnel of local, regional and national authorities working in the field of coastal landscape management may benefit from school/university education and training. At university level in particular, environmental conservation, landscape analysis, assessment, architecture, protection, Environmental Impact Analysis, management and planning are already served at a number of European universities (Landscape-Europe 2017).

In a report assessing the status of education and training of landscape architects in Council of Europe member states, Sarlöv-Herlin (2012) presented a number of recommendations on related curricula and educational structures, that include amongst others, a good theoretical knowledge and practical skills in landscape analysis and assessment understanding processes of landscape change and how to monitor them. This is of specific relevance to coastal scenery.

Lisitzin (2012) recommends that training for cultural landscape management should address a theoretical base for management and a number of methodologies that are:

- Inventory-based and analytical, allowing identification of natural and cultural heritage and assessment of the latter's local community value.
- Enable a participatory and integrated approach to management and planning.
- Representative of traditional economies and supportive of economic development.

In addressing management guidelines for IUCN Category V protected areas (Protected Landscapes/Seascapes) and more specifically, policies relating to tourism, public awareness, education, information and interpretation of such areas and scenery is an essential component of this. Phillips (2002) identified that these issues are closely related to tourism and their enjoyment of such areas; this is because tourism may contribute to an improved understanding and cognizance of conservation needs of such sites, among themselves and locals alike. However, Phillips (2002) also recognized that related-education needs of locals and visitors are different. Threatened with a potential underrating of their culture and traditions, local inhabitants require (at both school level and adult community learning and training initiatives), a nurturing of their awareness of their landscape and culture and its importance. On the other hand, tourists require an increased ability to recognize and respect the importance of a site and its inhabitants. In this context, they require not only simple operational facts about the protected area but also intrinsic knowledge of its cultural history and local ecology.

National and regional planning and management authorities should also ensure awareness raising and education of their staff on the availability and use of a range of tools and techniques designed for the evaluation of coastal scenery in its own right (see Sect. 7.3.5) and as an aid in the evaluation of coastal landscape.

7.3.5 Management of Coastal Landscape via the Application of Tools Evaluating Coastal Scenery

Despite a surge during the late 1960s and the '70s in the development of more objective and quantitative landscape assessment techniques that could be repeated by different researchers in a variety of geographical regions and still result in comparable results (Robinson et al. 1976). Buhyoff and Riesenmann (1979) identified a continued difficulty in the elaboration of quantitative assessment techniques for scenic impacts. The subsequent and further evolution of techniques assessing landscape scenery is extensively reviewed in Chap. 3. Since the term 'visual quality' has often been used as a reflection of beauty and projecting a level of objectivity (Jacques 1980) the evaluation of scenery via objective methodologies such as the CSES technique described in Chap. 4, may be considered as imperative for an effective reflection of the aesthetic influence of management actions (Buhyoff et al. 1994).

Landscape evaluation has been described by Unwin (1975, 130–131) as being composed of three phases:

- Landscape measurement: an objective description and classification of the landscape which produces an inventory of what actually exists with no consideration of 'scoring' on a qualitative basis.
- Landscape preference/value measurement: an investigation and measurement of value judgements or preferences for the visual landscape. Within this phase are included perception problems relating to the nature of the landscape images for which qualitative assessments are to be made.
- Landscape evaluation:the assessment of the quality of the objective visual landscape in terms of individual or societal preferences for different landscape types.

The above reflects the importance of the visual landscape and of techniques used to assess and evaluate coastal scenery to overall coastal landscape evaluation. In evaluating the U.K. Heritage Coast for example, scenery has been described as a prime determinant (Williams 1987, 2002). A review of several scenic evaluation tools is presented in Chap. 3 and a detailed description of the Coastal Scenic Evaluation (CSES) technique (Ergin et al. 2002, 2003, 2004, 2006) and several examples of worked evaluations using the CSES technique are presented in Chaps. 4 and 6. These chapters clearly reflect the potential of identifying low scoring physical features and human parameters that influence coastal scenery, that via targeted intervention and appropriate management may be upgraded to achieve improved quality scenery. This technique also demonstrates a potential to predict impacts of proposed future development on the quality of coastal scenery by considering their influence on the various parameters used for scenic rating. Scenic rating by such systems, are also useful to determine choice of management strategies (i.e. choosing that strategy which will result in, amongst others, improved scenic quality evaluations). In this manner, such scenic evaluation systems may guide optimum conservation, management and land use planning strategies of assessed sites.

In many conflict situations (e.g. with agriculture), scenery has been shown to take precedence and the concept has spread overseas, e.g. India and South Africa (Williams 1987). Leopold (1969) inventoried a checklist of parameters with regard to their suitability for addressing the aesthetic aspects of scenery in order to reduce subjectivity, so that, *'the results promise to be a useful, new kind of basic data needed in many planning and decision-making circumstances.'* (Leopold 1969, 1). Barbosa de Araujo and Costa (2008, 1448) indicated that in Brazil, landscape was probably highly rated as an attribute in visitor's choice and *'the maintenance of the landscape quality of beaches must be the main priority.'* Similarly, scenic destinations for tourism purposes now seem to be well ensconced in tourism literature, e.g. *'Landscape is most frequently exploited by the government and the media as advertisements of Pernambuco State tourist resources.'* (Barbosa de Araujo and Costa 2008, 1445). Utilizing the Geotraveler Tendency Behavior Scale, Boley and Nickerson (2013) showed that environmentally concerned tourists will specifically travel to an area for its scenic beauty.

In an assessment of coastal scenery that applied the CSES technique in 135 sites on the Caribbean coast of Colombia, Rangel-Buitrago et al. (2013, 2016) concluded that on the studied sites, increased management efforts should prioritise the upgrading of human parameters by reducing litter and indications of sewage, beach vegetation debris and enhancing beach nourishment works. Low scoring sites were generally identified to score low on many human parameters with associated problems such as high litter amount, sewage, noise disturbance, absence of buffer zones, poor skyline quality and intrusive utilities. The adverse impact of coastal erosion often resulting not only in a degraded environment but also the presence of many intrusive hard coastal protection structures was also cited as negatively influencing coastal scenic rating. In his considerations of landscape perception, scenery and the contribution of public participation to effective coastal management in Portugal Pereira da Silva (2006) commented on the effectiveness of this type of assessments.

Scenery monitoring can be a useful tool for management purposes because it may lead to the establishment of a strong legislation for better coastal management (Rangel-Buitrago et al. 2018). This activity can also stimulate a sense of environmental responsibility and enhance stakeholder participation in the activities related to the maintenance of environmental quality, such as, coastal cleanup programs. Raising public awareness of the overall quality of coast quality is the only guaranteed way to improve scenery (Storrier and McGlashan 2006).

7.4   Conclusions

Information presented in this chapter reveals the necessity of adequate coastal scenery management. A range of management approaches that may be used to this end were identified including opportunities presented by various landscape-related conventions and policiy strategies, direct management of coastal landscape constituent elements, management of proteced areas, education, training and increased public participation and, the application of tools evaluating coastal scenery. It also highlights the necessity of their inclusion in world environmental and political agendas. However, while the science is almost clear, corresponding integration needs to be enhanced for everyone's benefit. Currently, national policies and local actions taking an integrated and holistic approach to address the downward spiral of negative coastal scenery impacts, is in many instances, lacking.

Prioritizing the role of scenery across all coastal countries is urgent. The current conditions of climate change triggers a broad shift towards a global civilization that could be sustainable if Homo sapiens seizes the opportunities and starts acting intelligently. World countries, especially those that depend of coastal tourism, must try to manage scenery with fresh and innovative approaches, and at the same time integrate them into traditional coastal management processes such as ICZM.

Management is a complicated process that requires a holistic view to finding practical solutions that many times go beyond a national issue (Williams and Micallef 2009; Cooper and Pilkey 2012). Currently, coastal scenery management has become a worldwide imperative because for every day that passes, landscape degradation becomes more intricate and the solution by means of investment of large amounts of money is not always the best answer.

Flexible strategies and not reactive measures should be adopted to strengthen management and thus improve the coastal scenery quality. For example, Algae blooms seldom occur, but management plans must be updated and in force to deal with these occurrences. For this kind of event, it might be necessary to incorporate beach and river basin management.

Scenery management should be based on strategies that allow an inherent ability of the coast to recuperate from the impacts induced by humans while maintaining the functions fulfilled by the coastal system over the medium and long term. A short-term vision does not allow the above mentioned to occur. For this to take place, it is necessary to consider the different typologies and their related processes involved to define the strategy to be implemented (Williams and Micallef 2009; Rangel-Buitrago et al. 2018). The Coastal Scenic Evaluation System advocated in Chap. 4 togeter with the aplication examples given in Chap. 6 of this book presents an effective and easy to use management tool that exemplifies this point.

References

Abdulla A, Obura D, Bertzky B (2014) Marine world heritage: creating a globally more balanced and representative list. Aqua Conserv Mar Freshw Ecosyst 24(S2):59–74

Agnoletti M, Santoro A, Gardin L (2013) Assessing the integrity of the historical landscapes. Three case studies in some terraced areas. In: Agnoletti M (ed) Italian historical rural landscapes cultural values for the environment and rural development. Springer, Amsterdam, pp 89–130

Alexander D (1985) Environmental management. Springer, New York

Alves FL, Silva CP, Pinto P (2007) The assessment of coastal zone development at a regional level—the case study of the Portuguese Central Area. J Coast Res 50:72–76

Amir S, Gidalizon E (1990) Expert based method for the evaluation of visual absorption capacity of the landscape. J Environ Econ Manag 30:251–163

Anderson PF (1979) Analysis of landscape character for visual resource management. In: Conference on applied techniques for analysis and management of the visual resource. Incline Village, Nevada, pp 23–25

Anderson JR, Hardy EE, Roach JT, Witmer RE (1976) Land use and land cover classification system for use with remote sensor data. Geological survey professional paper 964. U.S. Geological Survey Circular 671. United States Government Printing Office, Washington, DC

Arianoutsou M, Delipetrou P, Vilà M, Dimitrakopoulos PG, Celesti-Grapow L, Wardell-Johnson G, Henderson L, Fuentes N, Ugarte-Mendes E, Rundel PW (2013) Comparative patterns of plant invasions in the Mediterranean biome. PLoS One 8(11):e79174

Aston Centre for Europe (2013) The relationship between desertification and climate change in the Mediterranean. European Committee of the Regions, London

Ayad YM (2005) Remote sensing and GIS in modeling visual landscape change: a case study of the northwestern arid coast of Egypt. Lands Urban Plan 73(4):307–325

Bajocco S, De Angelis A, Perini L, Ferrara A, Salvati L (2012) The impact of land use/land cover changes on land degradation dynamics: a Mediterranean Case Study. Environ Manag 49(5):980–989

BaltCoast (2005) The role of spatial planning in integrated coastal zone management – findings and recommendations from the INTERREG III. BaltCoast, Riga

Barrachina M (2007) The effects of land use change on landscape: the case of VallFosca (Catalan Pyrenees). In: Proceedings of man in the landscape across frontiers. IGU –LUCC Central Europe conference, pp 20–32

Barbosa de Araújo MC, Ferreira da Costa M (2008) Environmental quality indicators for recreational beaches classification. J Coast Res 24(6):1439–1449

Barton ME (1998) Geotechnical problems with the maintenance of geological exposures in clay cliffs subject toreduced erosion rates. In: Hooke J (ed) Coastal defence and Earth science conservation. Geological Society, London, pp 32–45

Beder S (2002) Economy and environment: competitors or partners? Pac Ecol 3:50–56

Blue Solutions (2017) Ecosystem-based coastal and marine spatial management. Blue Solutions, Berlin

Boley B, Nickerson NP (2013) Profiling geotravelers: an a priori segmentation identifying and defining sustainable travelers using the Geotraveler Tendency Scale (GTS). Jour Sus Tourm 21(2):314–330

Brewster A, Byrd BF, Reddy SN (2003) Cultural landscapes of coastal foragers: an example of GIS and drainage catchment analysis from Southern California. J Archaeol Sci 1:47–60

Brunori E, Salvati L, Mancinelli R, Smiraglia D, Biasi R (2017) Multi-temporal land use and cover changing analysis: the environmental impact in Mediterranean area. Int J Sust Dev World 24(3):276–288

Buhyoff GJ, Riesenmann MF (1979) Experimental manipulation of dimensionality in landscape preference judgments: a quantitative validation. Leis Sci 2:221–238

Buhyoff GJ, Miller PA, Roach JW, Zhou D, Fuller LG (1994) An AI methodology for landscape visual assessments. AI Appl 8:1–13

Burger J (2000) Landscapes, tourism, and conservation. Sci Total Environ 249(1–3):39–49

Burke SM, Thornes JB (2004) A thematic review of EU Mediterranean desertification research in Frameworks III and IV. Adv Environ Monit Model 1:1–14

Busquets Fàbregas J, Cortina Ramos A (2017) Management of the territory: landscape management as a process. In: Council of Europe, landscape dimensions. Reflections and proposals for the implementation of the European landscape convention. Council of Europe Publishing, Strasbourg

Cammelli C, Jackson NL, Nordstrom KF, Pranzini E (2006) Assessment of a gravel-nourishment project fronting a seawall at Marina di Pisa, Italy. J Coast Res 39:770–775

Carpaneto G, Paola G, Peccenini S (2003) Introduction. In: Minelli A (ed) The Mediterranean maquis – evergreen coastal formations, Italian habitats. Italian Ministry of the Environment and Territory Protection, Comune di Udine, pp 81–88

Cendrero A, Fischer DW (1997) A procedure for assessing the environmental quality of coastal areas for planning and management. J Coast Res 13(3):732–744

Colombini I, Chelazzi L (2010) Evolution, impacts and management of the Wetlands of the Grosseto Plain, Italy. In: Scapini F, Ciampi G (eds) Coastal water bodies-nature and culture conflicts in the Mediterranean. Springer, Amsterdam, pp 167–175

Conrad E (2010) People and landscape... Coming in from the cold. In: Conrad E, Cassar LF (eds) Perspectives on landscapes institute of earth systems. University of Malta, La Valeta, pp 23–31

Conservation International (2016) Mediterranean basin. Protecting nature's hotspots for people and posterity. Critical ecosystem partnership fund. Conservation International, Rome

Cooper A, Pilkey OH (2012) Pitfalls of shoreline stabilization – selected case studies. Springer, Dordrecht

Council of Europe (2000) Landscape convention. Council of Europe Publishing, Firenze

Council of Europe (2012) Landscape facets: reflections and proposals for the implementation of the European landscape convention. Council of Europe Publishing, Strasbourg

Coventry-Solihull-Warwickshire (1971) A strategy for the sub region, supplementary report 4 – evaluation. GVA Grimley, London

Cullingworth B, Nadin V (2006) Town and country planning in the UK. Routledge, London

de Groot RS (1992) Functions of nature. Wolters-Noordhoff, Groningen

DEFRA (2006) Shoreline management plan guidance volume 2: procedures. Department for Environment, Food and Rural Affairs, U.K. Government, London

Denier L, Scherr S, Shames S, Chatterton P, Hovani L, Stam N (2015) The little sustainable landscapes book. Oxford, London

Di Gregorio A, Jaffrain G, Weber JL (2011) Land cover classification for ecosystem accounting. United Nations Statistical Division, London

Dodouras S, Sorotou A (2011) Sustainable management of Mediterranean wetland landscapes. Int J Environ Cult Econ Soc Sustain 7(6):271–284

Doody P, Ferreria M, Lombardo S, Lucius I, Misdorp R, Niesing H, Salman A, Smallengange M (2004) Living with coastal erosion in Europe: sediment and space for sustainability; results from the eurosion study. Office for the Official Publications of the European Communities, Luxembourg

Dorset and Devon County Council (2009) Dorset and East Devon coast world heritage site management plan 2014–2019, Statement on the boundaries of the Site, and the World Heritage interests within them. Dorset and Devon County Council, Devon

Dudley N (2008) Guidelines for applying protected area management categories. IUCN, Gland

Dudley N, Schlaepfer R, Jackson W, Jeanrenaud JP, Stolton S (2006) Forest quality: assessing forests at a landscape scale. Earthscan, London

EC (2016) Management of Natura 2000 sites: best practice. EC, London

Education Services Australia (2013) Landscapes and landforms. GeogSpace. Australian Government Department of Education, Sidney

Ehler C, Douvere F (2009) Marine spatial planning: a step-by-step approach toward ecosystem-based management. UNESCO-Inter-governmental Oceanographic Commission and Man and the Biosphere Programme, Paris

Ergin A, Karakaya ST, Micallef A, Williams AT (2002) An innovative approach to coastal scenic evaluation. In: Ozhan E (ed) Proceedings of the international MEDCOAST workshop on beach management in the Mediterranean and the Black Sea: dynamics, regeneration, ecology and management. MEDCOAST, Kusadasi, pp 215–226

Ergin A, Williams AT, Micallef A (2003) Coastal scenic assessment at selected areas: a final project report. British Council, Ankara

Ergin A, Karaesmen E, Micallef A, Williams AT (2004) A new methodology for evaluating coastal scenery: fuzzy logic systems. Area 36:367–386

Ergin A, Williams AT, Micallef A (2006) Coastal scenery: appreciation and evaluation. J Coast Res 22(4):958–964

Estrada-Carmona N, Hart AK, DeClerck FAJ, Harvey CA, Milder JC (2014) Integrated landscape management for agriculture, rural livelihoods, and ecosystem conservation: an assessment of experience from Latin America and the Caribbean. Landsc Urban Plan 129:1–11

European Commission (2012) Integrated coastal zone management. OURCOAST – outcomes and lessons learned. Publications Office of the European Union, Luxembourg

European Commission (2016) Integrated coastal management. Publications Office of the European Union, Luxembourg

European Commission (2017) Commission welcomes Parliament's adoption of maritime spatial planning legislation. Publications Office of the European Union, Luxembourg

European Environmental Agency (2006) The changing faces of Europe's coastal areas. European Environment Agency, Report, Copenhagen

EUROSTAT (2017) Land cover, land use and landscape. Eurostat, Paris

Felter E, Morris M (2016) Coastal zone management. American Planning Association Planning Advisory Service, Washington, DC

Feranec J, Jaffrain G, Soukup T, Hazeu G (2010) Determining changes and flows in European landscapes 1990–2000 using CORINE land cover data. Appl Geogr 30:19–35

Geist HJ, Lambin EF (2004) Dynamic causal patterns of desertification. Bioscience 54:817–829

Hepcan S, Coskun Hepcan C, Kilicaslan C, Ozkan MB, Kocan N (2013) Analyzing landscape change and urban sprawl in a Mediterranean coastal landscape: a case study from Izmir, Turkey. J Coast Res 29(2):301–310

Hughes JD (2011) Ancient deforestation revisited. J Hist Biol 44(1):43–57

Hull RB, Revell GRB (1989) Issues in sampling landscapes for visual quality assessments. Landsc Urban Plan 17:323–330

IUCN (1994) Guidelines for protected area management categories. IUCN, Gland

Jacques DL (1980) Landscape appraisal: the case for a subjective theory. J Environ Manag 10:107–113

James Hutton Institute (2014) Review of existing methods of landscape assessment and evaluation. James Hutton Institute Publications, London

Jansen LJM (2004) Global land cover harmonisation: report of the UNEP/FAO expert consultation on strategies for land cover mapping and monitoring. In: Groom G (ed) Developments in strategic landscape monitoring for the Nordic countries. Nordic Council of Ministers of Environment, Oslo, pp 75–87

Jansen LJM, Di Gregorio A (2004) Land cover classification system: basic concepts, main software functions and overview of the "land system" approach. In: Groom G (ed) Developments in strategic landscape monitoring for the Nordic countries. Nordic Council of Ministers of Environment, Oslo, pp 64–73

ISPRA (2016) Guidelines for environmental studies related to the construction of coastal defence works

Komar PD, McDougal WG (1988) Coastal erosion and engineering structures: the Oregon experience. J Coast Res 4:77–92

Kuleshova M, Semenova T (2012) Landscape and ethics. In: Landscape facets – reflections and proposals for the implementation of the European landscape convention. Council of Europe Publishing, London, pp 306–325

La Mela Veca DS, Cullotta S, Sferlazza S, Maetzke FG (2016) Anthropogenic influences in land use/land cover changes in Mediterranean forest landscapes in Sicily. Land 5(1):3–13

Land Use Consultants (2017) Scottish natural heritage provides clear focus for Scotland's coastal landscapes with LUC input. Land Use UK, London

Landscape-Europe (2017) Education. Europe Publishing, London

Leopold LB (1969) Landscape esthetics: how to quantify the scenic of a river valley. Nat Hist 78(8):36–45

Letcher T (2015) Climate change – observed impacts on planet earth. Elsevier, New York

Lino CF, Britto de Moraes M (2004) Protecting landscapes and seascapes: experience from coastal regions of Brazil. In: Mitchell N, Beresford M, Brown J (eds) The protected landscape approach: linking nature, culture and community. World Conservation Union, London, pp 268–275

Linton DL (1968) The assessment of scenery as a natural resource. Scott Geogr Mag 84:219–238

Lisitzin K (2012) Capacity building: professional development and training. Council of Europe Publishing, Strasbourg

Masselink G, Hughes MG, Knight J (2014) Introduction to coastal processes and geomorphology. Routledge, Abingdon

Matar DA, Anthony BP (2017) UNESCO biosphere reserve management evaluation: where do we stand and what's next? Int J UNESCO Biosph Reserve 1(1):25–45

May V (2015) Coastal cliff conservation and management: the Dorset and East Devon coast world heritage site. J Coast Conserv 19(6):821–829

Meiner AJ (2010) Integrated maritime policy for the European Union – consolidating coastal and marine information to support maritime spatial planning. J Coast Conserv 14:1–11

Micallef A, Cassar M (2001) An environmental impact statement on the proposed beach nourishment project at Bajja ta' San Gorg, San Giljan, Malta. Malta Ministry for Tourism, St. Julians

Micallef A, Williams AT (2003) Application of function analysis to bathing areas in the Maltese Islands. J Coast Res 9:147–158

Micallef S, Micallef A, Galdies C (2017) Application of the Coastal Hazard Wheel to assess erosion on the Maltese coast. Ocean Coast Manage https://doi.org/10.1016/j.ocecoaman.2017.06.005

Mitchell N, Rössler M, Tricaud PM (2009) World heritage cultural landscapes a handbook for conservation and management. UNESCO, Paris

Mulligan M, Burke S, Ogilvie A (2016) Much more than simply "desertification": understanding agricultural sustainability and change in the Mediterranean. In: Behnke R, Mortimore M (eds) The end of desertification? Springer Earth System Sciences, Berlin, pp 427–450

Natural England (2009a) Guidelines for implementing the European landscape convention. Part 1: what does it mean for your organisation? Natural England, London

Natural England (2009b) Guidelines for implementing the European landscape convention. Part 2: Integrating the intent of the ELC into plans, policies and strategies. Natural England, London

Niemeyer HD, Beaufort G, Mayerle R, Monbaliu J, Townend I, ToxvigMadsen H, de Vriend H, Wurpts A (2016) Socio-economic impacts—coastal protection. In: Quante M, Colijn F (eds) North sea region climate change assessment. Regional climate studies. Springer, New York, pp 457–474

NOAA (2017) C-CAP land cover atlas. NOAA Office for Coastal Management. NOAA, St Peterbourg

ODNR (2017a) Ohio coastal atlas. ODNR Office of Coastal Management, Cleveland

ODNR (2017b) Chapter 4: land cover and protected areas. In: ODNR (ed) Ohio coastal atlas. 2nd ed. ODNR Office of Coastal Management, Cleveland

OECS (2000) Role of OECS in Environmental Management. OECS, Castries

Olsen SB, Tobey J, Kerr M (1997) A common framework for learning from ICM experience. Ocean Coast Manag 37:55–174

Pal M, Mather PM (2003) An assessment of the effectiveness of decision tree methods for land cover classification. Remote Sens Environ 86:554–565

Pereira da Silva C (2006) Landscape perception and coastal management: a methodology to encourage public participation. J Coast Res 39:930–934

Phillips A (2002) Management guidelines for IUCN Category V protected areas: protected landscapes/seascapes. Best Practice Protected Area Guidelines Series. 9. IUCN, Gland

Pilkey OH, Pilkey-Jarvis L, Pilkey KC (2016) Retreat from the sea: hard decisions in an age of global change. Columbia University Press, New York

Pranzini E, Williams AT (2013) Coastal erosion and protection in Europe. Routledge/Earthscan, London

Ramsar Convention Secretariat (2010) Managing wetlands: frameworks for managing wetlands of international importance and other wetland sites. Ramsar handbooks for the wise use of wetlands. Ramsar Convention Secretariat, Gland

Rangel-Buitrago N, Anfuso G, Correa I, Ergin A, Williams AT (2013) Assessing and managing scenery of the Caribbean Coast of Colombia. Tour Manag 35:41–58

Rangel-Buitrago N, Anfuso G, Ergin A, Williams AT (2015) Assessing and managing the coastal scenery: blue solutions from Latin America and the Wider Caribbean. GTZ, Berlin

Rangel-Buitrago N, Gracia A, Anfuso G, Ergin A, Williams A (2016) Evaluación de las características paisajísticas mediante la lógica matemática en la zona central de la costa Caribe Colombiana. Études Car (33–34):1–12

Rangel-Buitrago N, Williams AT, Anfuso G (2018) Hard protection structures as a principal coastal erosion management strategy along the Caribbean coast of Colombia. A chronicle of pitfalls. Ocean Coast Manag 156:58–75. https://doi.org/10.1016/j.ocecoaman.2017.04.006

Reed J, Deaki E, Sunderland T (2015) What are 'integrated landscape approaches' and how effectively have they been implemented in the tropics: a systematic map protocol. Environ Evid 4:2–5

Robinson DG, Laurie IC, Wager JF, Traill AL (1976) Landscape evaluation: report to the countryside commission for England and Wales. University of Manchester, Manchester

Romulus G (2004) Protected landscapes and seascapes and their relevance to small island developing states in the Caribbean: the case of Saint Lucia. In: Mitchell N, Beresford M, Brown J (eds) The protected landscape approach: linking nature, culture and community. World Conservation Union, Saint Lucia, pp 176–186

Rösch V, Tscharntke T, Scherber C, Batáry P (2013) Landscape composition, connectivity and fragment size drive effects of grassland fragmentation on insect communities. J Appl Ecol 50(2):387–394

Rössler M (2000) Landscape stewardship: new directions in conservation of nature and culture. George Wright Forum 17(1):27–34

Rössler M (2004) UNESCO world heritage centre background document on UNESCO world heritage cultural landscapes. FAO Workshop and Steering Committee Meeting of the GIAHS, Rome

Saeidi S, Mohammadzadeh M, Salmanmahiny A, Mirkarimi SH (2017) Performance evaluation of multiple methods for landscape aesthetic suitability mapping: a comparative study between multi-criteria evaluation, logistic regression and multi-layer perceptron neural network. Land Use Policy 67:1–12

Sarlöv-Herlin I (2012) Training of landscape architects. Council of Europe Publishing, Strasbourg

Scarascia-Mugnozza G, Oswald H, Piussi P, Radoglou K (2000) Forests of the Mediterranean region: gaps in knowledge and research needs. For Ecol Manag 132:97–109

Schaaf T, Clamote Rodrigues D (2016) Managing MIDAs: harmonising the management of multiinternationally designated areas: Ramsar sites, world heritage sites, biosphere reserves and UNESCO Global Geoparks. IUCN, Gland

Scottish Natural Heritage (2016) Geoparks. Scottish Natural Heritage, Glascow

Sluiter R, de Jong SM (2007) Spatial patterns of Mediterranean land abandonment and related land cover transitions. Landsc Ecol 22:559–576

Smith HD, Maes F, Stojanovic TA, Ballinger RC (2011) The integration of land and marine spatial planning. J Coast Conserv 15(2):291–303

Spalding MD (2012) Marine world heritage: towards a representative, balanced, and credible world heritage list. UNESCO World Heritage Center, Paris

Spiteri D, Scerri C, Valdramidis V (2015) The current situation for the water sources in the Maltese islands. Malta J Health Sci 25:22–25

Steers JA (1972) The sea coast. New Nat Ser 25:276–283

Storrier KL, McGlashan DJ (2006) Development and management of a coastal litter campaign: the voluntary coastal partnership approach. Mar Policy 30(2):189–196

Terkenli TS (2010) Understanding and analysing cultural landscapes. In: Conrad E, Cassar LF (eds) Perspectives on landscapes, institute of earth systems. University of Malta, Msida, pp 101–125

UNECE (2008) Spatial planning – key instrument for development and effective governance with special reference to countries in transition. United Nations economic Commission for Europe, Information Service, United Nations, Geneva

UNEP/MAP/PAP (2008) Protocol on integrated Coastal Zone management in the Mediterranean. Priority Actions Programme, Split

UNESCO (2016) Operational guidelines for the implementation of the world heritage convention. UNESCO Intergovernmental Committee for the Protection of the World Cultural and Natural Heritage. UNESCO, Paris

UNESCO (2017a) Biosphere reserves – learning sites for sustainable development. UNESCO, Paris

UNESCO (2017b) UNESCO Global Geoparks. UNESCO, Paris

UNESCO (2017c) Decision: CONF 208 VIII.C The Costiera Amalfitana (Italy). UNESCO, Paris

UNESCO World Heritage Centre (2017) World heritage list. UNESCO, Paris

UNFCCC (2007) Climate change: impacts, vulnerabilities and adaptation in developing countries. United Nations Framework Convention on Climate Change Secretariat Information Services, Paris

Unwin KI (1975) The relationship of observer and landscape in landscape evaluation. Trans Inst Br Geogr 66:130–133

van der Maarel E (1978) Ecological principles for physical planning. In: Holdgate MW, Woodman MJ (eds) The breakdown and restoration of ecosystems. Plenum Press, New York, pp 413–450

van der Maarel E (1979) Environmental management of coastal dunes in the Netherlands. In: Jefferies RL, Davy AJ (eds) Ecological processes in coastal environments. Blackwell, Oxford, pp 543–570

Verhoeven JTA, Setter TL (2010) Agricultural use of wetlands: opportunities and limitations. Ann Bot 105(1):155–163

VLIZ (2013) Policy-informing brief: EC consultation on the possible ways forward for maritime spatial planning and integrated coastal zone management in the EU. VLIZ, Ostend

Ward K, Snoberger N (2009) Assessment of landscape scenic quality in the Angelina national forest, Texas using GIS and high-resolution digital imagery. ASPRS/MAPP, San Antonio

Whitehouse R, Cooper N, Laeger S, Surendran S, Kermode T (2017) Predicting large-scale coastal morphological change: from fact to fiction. Delf, The Hague

Williams AT (1987) Coastal conservation policy development in England and Wales with special reference to the heritage coast concept. J Coast Res 31(1):99–106

Williams AT (2002) Coastal cliff engineering: case studies from Wales, UK. Coast Eng 6:44–59

Williams AT, Micallef A (2009) Beach management principles and practice. Earthscan, London

Williams AT, Rangel-Buitrago N, Pranzini E, Anfuso G (2018) The management of coastal erosion. Ocean Coast Manag 156:58–75. https://doi.org/10.1016/j.ocecoaman.2017.03.022

Wisconsin Department of Natural Resources (2014) Land cover and agricultural management definition within the upper Wisconsin river basin. Bureau of Science Services, Wisconsin

Zandmotor D (2017) The sand motor: successful driver of innovative coastal management. Delft, The Hague

Zube EH, Pitt DG, Anderson TW (1974) Perception and measurement of scenic resources in the Southern Connecticut River Valley. Institute for Management and Historic Environment, Amherst